THE OFFICIAL CB BOOK:

A CONSUMER'S GUIDE

THE OFFICIAL CB BOOK:
A CONSUMER'S GUIDE

Jon F. Thompson

BALLANTINE BOOKS · NEW YORK

For Carolyn,
Jonathan, and Matthew

Published in the United States by Ballantine Books, a division of Random House, Inc., New York and simultaneously in Canada by Ballantine Books of Canada, Ltd. ,Toronto Canada.

Library of Congress Catalog Card Number: 76-9762

ISBN 0-345-25239-X-395

Manufactured in the United States of America

First Edition: June, 1976

Designed by Gene Siegel

CONTENTS

INTRODUCTION

It was a media specialist named Marshall McLuhan who noted, about ten years ago, that "the medium is the message." McLuhan couldn't have known it at the time, but if there ever was anything that proved his point, it would have to be Citizens Band radio.

The airwaves used by CB signals were, just a few short years ago, the exclusive stamping ground of a relatively small group of radio hobbyists, sportsmen, and business people who used their sets for relatively narrow purposes, but all that has changed in the last two years. Citizens Band radio has been discovered by the press and, because of that, by you and me. Folks are buying CB sets left and right. They're buying them for use between the wife at home and the husband out in the car; they're buying them for the instant, inexpensive, convenient two-way communications they provide. On top of that, more hobbyists than ever are buying into Citizens Band most importantly because it's a heck of a lot of fun, but also because, while the sets are small and for the most part inexpensive, they still carry a considerable amount of sophistication. A CB radio lets you be where the action is, and puts you in touch with other folks who are as excited about the prospects raised by Citizens Band radio as you are.

There are lots of folks of both persuasions—call them "practicalists" and "hobbyists"—and there are boxcarloads of really fine equipment available. But when it comes to figuring out how to get the most for your money, how to buy the right set, how to buy the right antenna, and how to make it all work as it is supposed to work, the need for a nontechnical beginner's handbook on Citizens Band radio and its uses becomes strikingly apparent. Because so far there just hasn't been such a thing available—a fact you're already aware of if you've bummed from CB store to CB store in search of a book that will make it all clear to you.

This book is a simple, nontechnical attempt to help fill that gap, and the fact that it eventually got written is due in no small part to the help and guidance I received from people like Bill Cody of Pace Communications, Richard Everett of the Federal Communications Commission, Gordon West, Steve Roecker, and Leon Mandel.

So read on, learn, and enjoy yourself. Because that—enjoying yourself—is what a good part of Citizens Band radio is all about.

JFT

Reno, Nevada
March 3, 1976

1

WHAT CB RADIO IS, WHAT IT DOES:

An Overview

What you should know if you are interested in getting into Citizens Band radio is that you most definitely are not by yourself. There are these days, by the best estimates of the Federal Communications Commission, more than six million CB radio users. And the FCC ought to know, because it is the arm of the federal government charged with overseeing the airwaves used by CB radio. The FCC reports that about half of those six million people actually have CB licenses, and that the number of licensed CBers is growing rapidly, thanks at least partly to a liberalization of the rules and regulations that govern Citizens Band radio and its use. But we'll get a little deeper into that later.

What's important right now is to learn a few of the basics about CB radio: what it is, what it is not, what it can do and what it cannot do. Compared to other types of two-way radio communications, Citizens Band radio is very inexpensive, but even so, a small cash outlay is required. That being the case, you'll probably be relieved to learn that most observers of today's CB scene feel that the incredible growth of CB's popularity over the past two years is not the result of a mere fad, any more than FM car radios and stereo tape players in cars are fads. CB has become popular because the general public has begun discovering that it is a simple, inexpensive, reliable electrical appliance that provides wireless radiotelephone in a car or in a home. It's two-way radio for everyone, regardless of income or interest, and getting a license to operate a CB set is easy.

CB is in constant use by truckers, who use it to keep themselves company on long trips, to find diesel fuel when it is in short supply, and to spot traffic cops. It's in use by farmers, who can be plowing out in the south forty and still be in constant contact with the farmhouse. CB is in use by people who spend their free time in four-wheel-drive and recreational vehicles getting away from civilization but who still want to keep radio contact with it. And CB is going into private automobiles, in the city and in the country, at a lightning pace. If you've kept your eyes open you already know that, if for no other reason than the sudden proliferation of willowy CB antennas that seem to be growing up everywhere, from the trunks of cars, the tops of pickup trucks, and the peaks of suburban houses. Citizens Band radio is growing especially fast in trendy California.

As mentioned, Citizens Band radio can be considered to be not much more or less than an inexpensive wireless telephone with a medium-short range. Compared to ham radio equipment, which is costly, pumps out enormous amounts of power, and can be heard across oceans and continents, CB's advantages are many. Citizens Band is lightweight. The absolute maximum amount of power the FCC says can be pumped out of a CB transmitter is 4 watts; that relatively small amount of power is one of the main limitations on CB's range, so you'll not be using CB to talk to friends in Europe. But there is another side to the coin, which is that the lower power helps eliminate signals that stray off their assigned channels to muddy up signals being sent on adjacent channels. It's also one of the reasons the sets are inexpensive, since more powerful radio sets cost more to design and build.

The other main limitation to CB's range is the antenna, and we'll talk about that in detail in a later chapter. Suffice it to say now that the maximum allowable height for a CB antenna is 60 feet. Typical range for a simple base-station antenna mounted at that height could be as much as 30 miles or more, if you happen to live in the flatlands. Maximum range of most mobile units is usually approximately 15 miles, this because most mobile antennas, by their very nature, can't be as efficient as the larger base-station antennas.

The basic equipment involved in a CB set is simple enough:

Radio Spectrum

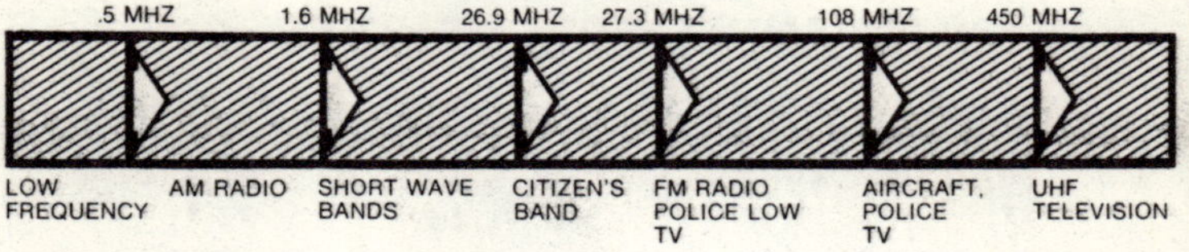

Figure 1-1. This a portion of the lower end of the radio wave spectrum. As you can see, Citizens Band radio uses a bit of the spectrum that falls between the AM and short wave radio bands on the low side and the FM radio and television bands on the high side. *Steve Roecker*

a combination radio transmitter/receiver called, aptly enough, a transceiver; an antenna; a bit of coaxial cable; and a simple device known as an SWR (standing wave ratio) meter, which will help you keep your antenna electrically matched to your transceiver, thus keeping your transceiver from burning its innards to a crisp.

How much you spend on this basic equipment is purely a function of how little or how much you feel you need in the way of extra features. All of today's CB sets must meet rigid standards set by the FCC, and by and large, the electronic circuitry of most sets is really rather good. So the performance of sets of different manufacturers should be more or less equal, though there are some important variations. We'll get to the specifics of those variations and how they affect set performance in a later chapter.

Depending on the features you think you want (and there are a lot of choices), and depending on whether or not you're looking for a mobile station for your car or truck or for a base station to plug into your living room or den, you can figure on spending anywhere between eighty dollars and eight hundred dollars just for the transceiver. Antenna prices range from less than twenty dollars to more than two hundred dollars, with the mobile antennas generally occupying the lower price ranges and the really sophisticated base-station antennas giving your wallet the heavier hits.

Discussion about radio and how it works can get pretty technical and difficult to understand, and we'll try to avoid that. A simple discussion of how a CB set works *is* called for, though, so hang in and follow along.

Citizens Band radio works on a very narrow portion of the

CITIZENS BAND RADIO CHANNEL FREQUENCIES			
Channel	**Frequency***	**Channel**	**Frequency***
1	26.965	13	27.115
2	26.975	14	27.125
3	26.985	15	27.135
4	27.005	16	27.155
5	27.015	17	27.165
6	27.025	18	27.175
7	27.035	19	27.185
8	27.055	20	27.205
9	27.065	21	27.215
10	27.075	22	27.225
11	27.085	23	27.255
12	27.105		

*listed in megahertz

Figure 1-2.

radio spectrum (the radio airwaves, in other words) that occupies the area between 26.9 and 27.2 megahertz. Megahertz, by the way, is a term that means one million vibrations, or cycles, per second, so the CB airwaves range from 26.9 million cycles per second to 27.2 million cycles per second. The radio spectrum extends from below .5 megahertz, which is down where AM radio works, up to more than 300,000 megahertz, which is the portion of the airwaves used for communications from satellites. So you can see just how small a portion of the radio airwaves is used by CB.

Rather than employing a radio dial for tuning from one of CB's twenty-three channels to another, CB sets are crystal tuned. This means that a CB set uses electronic crystals that are designed to vibrate at different specific rates. So if you wish to broadcast, say, on channel 1, instead of groping around a complex tuning dial looking for 26.9 megahertz, you merely flip to channel 1 the same way you change channels on a television set and bring into play a combination of crystals designed to give you the exact frequency you need.

The FCC currently is examining the possibility of granting CB radio additional airwave space—as many as one hundred additional channels have been mentioned—but this move could be as long as five years off, though it could happen sooner, too. Just where the new band will be, and whether or not the move will render present CB equipment obsolete, is unclear as yet. What does seem clear, however, is that if the move is made, the radio equipment that will be required to use the new band could be considerably more expensive than current equipment. Chances are that the current CB band probably will not be abandoned, though it probably will become even more crowded than it is now. This one factor, more than any other, is likely to cause a slow but steady migration of CBers to any newly opened channels.

A major characteristic of the portion of the airwaves that contains the CB band is that the radio signals travel pretty much along line of sight. That means that if you can't *see* past it (large buildings, a mountain), your signal probably won't be able to *get* past it. And because of where CB falls in the airwaves, there is another interesting oddity. It's called "skip," and it means that under certain atmospheric conditions, waves at certain radio frequencies may bounce—or skip—along the lower layer of the earth's ionosphere to be heard by a receiver as far away as a thousand miles from the source of the signal. Use of the skip phenomenon for transmission is illegal, and if you're caught, you just might have some explaining to do to the FCC. Skip will be discussed further in Chapter 7.

Just how the FCC polices the airwaves, especially those assigned to Citizens Band radio, is of interest. The rules and regulations governing CB, which are contained in the FCC's Volume VI, Part 95, underwent some basic changes in the summer of 1975. The old rules reflected a different time and a different way of thinking about who should use CB radio and for what. They were fairly restrictive and unpopular. Flouting of those rules was widespread, for there was little chance of the FCC's ever getting the budget to adequately crack down on rule breakers. So in one swoop the FCC made the rules at once more palatable to CBers and much more enforceable. Under the revised rules you still can't use skip, you can't broadcast without a license, you must not swear, and when a conversation becomes 5 minutes long you must break off for 60 seconds.

With fewer rules to enforce, the FCC says it intends to enforce them. To this end it has forty-five field offices, or bureaus, scattered around the country. One of their purposes is enforcement. The field offices are equipped with special vehicles designed to monitor radio signals. The FCC doesn't talk much about them, but they're used by FCC technicians who cover an assigned territory. They may be vans or cars, and they carry equipment designed to pinpoint sources of illegal radio signals. The system works well, and the FCC doesn't hesitate to cite an illegal operator—which is the way it should be.

The rules revamping, by the way, made it much easier and cheaper to get a license. The fee for a five-year ticket was lowered from twenty dollars to four dollars. Before the rules change in July 1975, license applications were running at approximately sixty thousand each month. By January 1976, monthly license applications had jumped to four hundred thousand and are still climbing.

You see the result: interest in CB is widespread, and growing fast. CB specialty shops are springing up right and left, department stores are offering Citizens Band radios for sale, and dealers for Ford, Chrysler, and General Motors will install after-market CB sets in their new cars—and factory new-car CB installations are not very far away.

So if you're getting into CB, you're not alone, and with all those other CBers out there, the channels are going to get crowded. It's important for you, then, to be very careful in your selection of transceiver and antenna, so that you can send the strongest possible signal and receive even the weakest ones. That means buying the right equipment in the first place, setting it up so that it will provide maximum performance now and forever, and using it to the best advantage. The chapters that follow will help you get all this accomplished, but first what you have to do is apply for your Citizens Band license.

2

ON GETTING A LICENSE:

Or, Starting at Square One

Chances are that if you have purchased this book you have a serious interest in Citizens Band radio. You may even intend to invest at least a *little* cash in a CB set, and if so, you're probably interested in doing the whole thing right. Bravo; like anything else in life, the use of Citizens Band radio seems to yield the best returns if first, you know what you're getting into, and second, you don't try to paddle upstream.

If you've launched an interest in Citizens Band radio and are wondering just which way you should starting paddling, you might as well know right now that the very first thing you should think about doing is getting a Citizens Band license. You can't broadcast without one (well, you can, but that's highly illegal and just not very wise), and once your application is mailed it takes a while before you actually receive your license, so you might as well make your mind up to apply now and use the waiting time to learn all you can about CB radios and how they are used. Then you really will know what's what when you take your money CB shopping.

If you've ever known anyone involved in amateur radio, you probably know that the licensing procedures for that form of two-way communications are long and involved. A would-be ham, as amateur radio operators are called, is asked to take a lengthy test that not only examines his knowledge of the technical side of radio, but makes sure he knows, and can send, Morse code.

Not so with Citizens Band radio. The whole idea of CB is to open the airwaves to the greatest possible number of people. Which is not to say that CB operators enjoy complete freedom from governmental control, because they don't. But minimal control is exerted, and in just three areas: 1) the Federal Communication Commission's general control of the airwaves, to which anyone who uses any kind of radio equipment is subject; 2) type acceptance of CB equipment, which means that all CB equipment must meet certain specific performance standards set by the FCC; and 3) the licensing procedure. The first of those items has already been mentioned, and the second will be discussed in its turn; the last will be discussed here.

The first, and probably the most important, point that should be made is that there is no reason at all to broadcast on a Citizens Band radio without benefit of a license; indeed, there are several very good reasons why you should refrain from broadcasting until your license arrives in the mail. To broadcast unlicensed is a violation of law and can result in some rather disquieting penalties, including confiscation of your equipment and payment of a stiff fine. The most telling argument of all, however, is that since the 1975 liberalization of the rules governing communication by Citizens Band radio, licenses are extremely easy to get. There is a bit of waiting time involved, that's true, but the procedure is a snap.

The cost is low, just four dollars, and the license, once you get it, is good for five years. A little note here that might help point out just how radically the government's attitude toward CB radio has changed since the 1975 rules revamping: Before the change, the cost of a license was twenty dollars. In the period of time leading up to the rules change, partly because of outside pressure, the FCC began to realize that a twenty-dollar license fee didn't do much more than create millions of radio outlaws, people who just wouldn't—or couldn't—pay twenty dollars for a license. Yet CB is supposed to be accessible to the public, and a lower license fee seemed called for, so the amount required was reduced to four dollars.

On the whole the restrictions are few. You can't use a CB radio in any way contrary to federal, state, or local laws. You must be at least eighteen years of age. You must not be a representative of a foreign government, and you must have some sort of mailing address. The FCC has placed a few restrictions on the equipment involved (on things like antenna height and transmitter power) and has established certain rules of operation, so you'd best go into CB knowing that you do have some light fetters.

The above paragraph mentions that the license applicant must be at least eighteen years of age. That's a fact, but it doesn't mean that you have to be eighteen to use a CB set. Anyone of any age can broadcast on Citizens Band radio, because when the FCC grants a license, it licenses the installation and operation of radio equipment, not the operators of that equipment. So anyone of any age can use your CB station, as long as they're under your supervision and you're willing to assume full responsibility for their conduct.

So the first thing to do is obtain a license application form, known as FCC Form 505, three or four transmitter-identification stickers, and the set of FCC rules and regulations contained in Volume VI, Part 95. You'll find a copy of the FCC's Form 505 enclosed in the pages of this book. If someone has beaten you to it, you can obtain the license forms and identification stickers by writing to the Federal Communications Commission, Washington, DC 20554, or by contacting your local FCC field office, which will be listed in your telephone directory under "United States Government." Some manufacturers enclose license forms and ID stickers in the cartons with their radios.

Volume VI, Part 95 of the FCC rules and regulations can be obtained by sending $5.35 to the Superintendent of Documents, U.S. Government Printing Office, Washington, DC 20402. When you sign the license application you certify that you either now own or have on order a set of the rules and regulations (read the fine print on the application), so you might as well stop right here and mail off a check.

You should know that your $5.35 buys what amounts to a long-term subscription to the FCC rules that govern CB radio and its use. The initial fee entitles you to receive what are called transmittal sheets, periodic updates to the rules and regulations made by the FCC. So when your copy of Part 95 arrives in the mail, get yourself a ring binder to keep all the rules and updates in. You can keep the binder right by your base station, or in a handy spot in a bookcase if you're interested only in a mobile unit. Then if you have a question you can flip to the page you need and find an answer.

The license application consists of just one form, and filling it out is as simple as getting a speeding ticket. The instructions are straightforward. You are asked to print clearly or use a typewriter, to please remember to date and sign the application, and to please not send cash.

Item one asks you to write down your first name, middle initial, and last name. Item two requests the month, day, and year of your birth. Three asks for the name of your business, and is to be completed only if you intend to use your CB set in the course of your business. Four asks you to list your street number and street name. You are referred to item eight if you have a post office box or rural free delivery number (RFD). Eight asks merely for the street address at which your main CB unit will spend most of its time.

Item five supplies space for you to write in the name of your city; six and seven request your state and zip code. If you filled out item eight you will have noticed that items nine and ten also call for your city and state. Fill them out only if you have used item eight. In every other case you must supply a response where there is a number and a box, or boxes, for you to write in.

Item eleven tries to narrow down just what kind of applicant you are: an individual? a business partnership? a single individual in business? an association? a corporation? a branch of local government? If none of these categories fits you, you are obliged to check the "Other" box, and to specify just what category you feel you fit into.

Item twelve asks you if this application is for a new license, to renew an existing license, or to increase the number of transmitters you specified that you would be using in a previous license application. If it's a renewal or an increase in licensed transmitters you're after, you already have a license, so there is a series of seven little boxes in which you are to write your current license number. If you don't have a license yet, just ignore this item.

Item thirteen contains two boxes, only one of which you should check. If you're looking for a class-C license, a non-voice, remote-control-of-models license, check the first one. But chances are you're not too interested in that, and are looking instead for a class-D voice license, in which case you should check the second box.

Item fourteen asks how many transceivers you intend to use, and here you will notice that the FCC has given you a great deal of latitude. Instead of asking you to come up with an arbitrary decision right now, you can check a box for one to five transceivers—which will be plenty for most individuals—or for six to fifteen—fine for a small business with a little fleet of cars or trucks—or for sixteen or for more. If you check this last box, there is a space just to the right of it in which you must specify how many transceivers you intend to operate. Fill it in. It's a good idea to do a little advance planning here and ask to have the maximum number of transceivers you're likely to use licensed. That will save you from going through this whole procedure again when it comes time to ask for a license revision.

You should read item fifteen before you sign the application, because when you sign the application, you verify that you have indeed read it. It certifies merely that you are neither a foreign government nor a representative thereof, that you either own or have ordered a copy of Part 95, that you will operate your equipment in full compliance with existing laws, that you will use it for no purpose against federal, state, or local laws, that you will not run your set with greater power than authorized, and that you understand the authority of the United States Government to govern the Citizens Band airwaves.

Item sixteen is where you sign your name, and seventeen is where you write in the date.

Now, there's still an unfilled-in portion of paper on the bottom of the application. Near the left-hand border, in boldface type, are the words "ORDER FORM." That's what it is. Use it to order a copy of Part 95 for yourself if you haven't already done so. Everything you need to know in order to fill out the form properly is right there *on* the form, so just go ahead and read the thing and do it; don't forget to include your check for $5.35 for this.

So. You've mailed your application, say, two weeks ago, and now you're waiting by your mailbox with bated breath for your license and call numbers. Don't do it. Processing of CB license applications is taking twelve weeks these days, with no relief in sight. According to the FCC, it should never take *longer* than twelve weeks, but you know how things go in the federal bureaucracy. Give your application an extra week, just for good measure.

One thing you should not do is get impatient and write (or call) to find out what the holdup is. Because that's what you'll create by doing so: a holdup. On your own license. Your application could be removed from the regular process chain so that it can be examined for the cause of the holdup you inquired about. Your best bet is simply to fill it out right the first time, and then be patient.

If you already have a radio, when your license finally does arrive you'll be ready to transmit. (You've just been listening all along, right? You should have been; there's nothing illegal about that.) But first you must post your license, your call numbers, and your name and address on your radio. Use the ID stickers you requested when you sent away for your license application and stick them to each transceiver you own. If a base station is what interests you, figure on having your license, or a copy of it, posted near the radio.

Just remember, if you move to a new address, if you change the number of transceivers specified on your license, or if your applicant category changes, you must file a new application accompanied by another four-dollar check. When your new, modified license arrives, send the old one to the FCC.

It's wise to get the license application filled out and mailed before you do anything else, especially before you go out and purchase a radio. Then, while you are waiting for the mailman to deliver your license, you'll have time to finish reading this book and understand what it will tell you about how to shop intelligently for equipment and how to use the equipment once you've bought it.

3

BEGINNINGS:

How CB Radio Began, When It Grew, and Why

Radio—both the kind we listen to for music and news and the two-way variety—just seems like it's always been with us. It's as familiar as a doorknob, and as homey as the smell of chicken frying. There isn't anyone these days who can't tie radio in some form or another to good times and places of days long past.

Well, radio has been with us for a while, and while it's a good lot older than television, its glamorous electronic media cousin, radio is a brash newcomer when compared to certain other methods man has devised to transmit information.

Radio can trace its origins to before the turn of the century, when Thomas A. Edison had a working telegraph device that could send messages over short distances without the use of wires. Radio was in experimental use by at least one of the nation's rail carriers by as early as 1915, and not long after that commercial radio began making its way into nearly every home.

There was no overseer of the radio airwaves in those early years. Anyone with the know-how and the money could put up a radio station, and if the businessman on the other side of town already had installed a commercial radio transmitter, all you had to do was build a more powerful one and drown him out. This is pretty much the way things went during radio's formative years; there were few private radio communications, and what commercial radio broadcasting there was was totally uncontrolled. Broadcasts interfered with one another considerably. True enough, mobile radio telephones had made an appearance by the late '20s, but they were heavy and costly, and man's knowledge about what radio was and how it worked was relatively unsophisticated. So development of two-way communications did not exactly proceed at a wildfire pace during those early years.

Federal regulation of the airwaves, covering both commercial and two-way uses, was initiated in 1927 by an act of Congress. The name of that first regulatory body was the Federal Radio Commission, and in this first incarnation federal control of the airwaves was carried out under the umbrella of the Department of Commerce. Some sort of control was necessary, in the eyes of the federal government, because the various radio stations around the country apparently did not care to regulate themselves.

(Involved with the Federal Radio Commission, by the way, was the pre-Presidential Herbert Hoover. Both Hoover and his son were considerably interested by the goings-on of this fledgling form of communications, and eventually both became radio hams.)

By 1933 it had become apparent to the federal government that radio in all its forms was growing far faster than anyone had anticipated. Not only that, but preliminary experiments with a whole new form of electronic wireless communications was producing some exciting results. The name of that new form of communications was television, and it seemed apparent from the outset that some form of supervision that would encompass both these media was called for.

As a result, the Congress passed the Communications Act of 1934. The act created the Federal Communications Commission, a body designed to oversee the growth and operations of electronic communications in the United States.

Two-way radio at this time was still pretty much the exclusive province of the military and of commercial, industrial, and governmental interests, and the relatively high costs of the radio equipment required for two-way communications was most of the reason. Another reason radio technology progressed rather slowly was, to put it simply, that nobody was spending the money to make it do otherwise. The gathering clouds of war soon changed that picture. With the coming of World War II, the development of more and more sophisticated radio technology suddenly was top priority because of the armed struggle taking place in Europe and in the Pacific. By the time the war ended, in 1945, a considerable amount of knowledge about electronic communications had been amassed, and Joe Average, the man on the street, was about to benefit from it.

In 1945 a series of Congressional hearings called the Land Mobile Frequency Allocation Hearings resulted in a growing determination that the public owned the airwaves and thus had a right to use them. These hearings had to do with the use of mobile radio telephones for business and commercial use, and also for use by the average citizen. Existing frequency allocations were thoroughly shuffled, and the space in the 460–470 megahertz band was allocated to the use of something named the Citizens Radio Service; this space was best suited to the performance characteristics of the two-way radios easily available during this time.

The hearings resulted in a document known as Docket 6651, which proposed the creation of the Citizens Radio Service, and two years later, in 1947—the wheels of the federal government ground as slowly then as they do now—the first set of rules had been generated and made public.

Two classes of Citizens Radio Service licenses had been created, class A and class B. Both had more or less the same purpose, but class A allowed up to 60 watts input, while class B was limited to 5 watts input.

Public reaction to this new form of private radio communications was, in a word, underwhelming, and for two very good reasons: The cost of the equipment needed to work the 460–470 megahertz band was still very high; and, because of the CRS's location in the radio spectrum, communications were very much line-of-sight, and thus range was very limited. These radios worked fine if they had a really high-power transmitter to push their signals and really exotic antenna systems to radiate those signals, but as far as the Citizens Radio Service was concerned, neither high power nor exotic antennas was very practical. Some of the problems created by the short-range characteristics could be mitigated by getting into class A instead of class B, but the A stations were much more expensive than the B stations, and needed considerably more maintenance. (Class C, by the way, also was approved at about this time. Its purpose was then and is now regulation of radio-controlled appliances and the like—everything from garage-door openers to model airplanes.)

This is not to say that the 11-meter band (another name for the frequencies around 27 megahertz) now enjoyed by Citizens Band radio was ignored. The war had opened up an interesting new research field, later used by industrial, scientific, and medical interests. It was called radio diathermy, and had to do with the fact that certain radio frequencies produced heat when they came into contact with certain substances. This sort of radio activity became known as ISM (Industrial, Scientific, Medical) and was assigned to 27.1 megahertz, smack where Citizens Band radio lives today. This frequency range was barred by the FCC from use for communications, and, on the surface, at least, this seemed sensible, because ISM users produced the desired effects right at their transmitters; interstation contact was not the point of this research.

Enter a group of radio hams who petitioned the FCC for use of the 11-meter band. They said that if such use was granted, they would happily put up with whatever interference ISM activities caused, and would cause no interference themselves. The FCC acceded to their petition, and the 11-meter band was in business.

Meanwhile, on the 460–470 megahertz area of the radio spectrum, it was business as usual—which is to say, slow. But around 1948 something happened that changed all that: An FCC commissioner named Jett wrote an article for the *Saturday Evening Post,* in which he sang the praises of the newly created Citizens Radio Service and told what it was, what it did, who could use it, and how they could benefit from it. The *Post* enjoyed a wide distribution in those days, and Commissioner Jett's story seemed to be just the boost the Citizens Radio Service needed. Shortly after the article appeared, license applications began a slow but steady climb—so slow that it took sixteen years for the FCC to get its millionth CRS license application.

Eventually the FCC realized that broadening access to CRS just might be to the advantage of everybody, and 1958 saw sweeping revisions, embodied in Docket 11994. Docket 11994 created a fourth class of Citizens Radio Service usage, class D, and placed it squarely in the eleven-meter band. There was a small outcry from the ham radio people, but, undaunted, the FCC went ahead with its plan. It seemed apparent that placing class D in the 11-meter band would help nullify the two biggest problems of the Citizens Radio Service: the relative scarcity and high cost of equipment, and the touchy, short-range frequencies generated by transmitters working 460–470 megahertz.

In the beginning of this reconstituted service, the FCC practiced relatively little regulation, but still reaction from the public was lukewarm. Sportsmen were using class D, as were construction and various professional people, but so far the general public—the people class D specifically was aimed at—just hadn't gotten the word. License applications hardly increased, and by January of 1959 they still numbered only slightly more than six hundred a month. But then a national electronics magazine got hold of the Citizens Band radio story. It gave details of how to build a simple CB set, and of the revised FCC rules, and with this media exposure Citizens Band radio suddenly took off. By May 1959 the number of monthly license applications had jumped to nearly six thousand. Two years later nearly two hundred thousand people had applied for licenses.

In those first few years after the 1958 revisions, people who used Citizens Band radio were relatively unfettered. The early rules were vague, as befitted the idea of making the media available to the widest possible group of people, and there just wasn't much regulation. The rules were general in nature, and there were many abuses. So in 1964 the FCC cracked down. It had become abundantly clear that some sort of regulation had become necessary, so the FCC amended its rules so that CB communications were tightly controlled. Tight, strict rules are fine under some circumstances, but in this instance all the FCC gained with its rules revisions was its very own enforcement nightmare. All sorts of people were broadcasting without licenses, all sorts of illegal equipment was being used to boost the power generated by class D stations, and all sorts of illegal modes of broadcasting were in everyday use by hundreds of thousands of people across the country. The FCC would have needed a small army to properly enforce all its rules, and it didn't have such an army. It wisely recognized that it had a bit of a problem, and eventually moved to ease some of the rules. That finally happened in 1975; the FCC's Docket 20120 was aimed at achieving a higher degree of voluntary compliance with the FCC's rules and regulations, and it all has worked remarkably well. The FCC says it took sixteen years to get its first million license applications and only a year to get the second million; these days nearly half a million applications roll into FCC headquarters each month. The easing of the

FCC's rules eliminated the prohibition on hobby use of Citizens Band radio, and concentrated on enforcement of licensing, station identification, use of proper frequencies, and some other points. It also made licenses easier and cheaper to get.

There was one more reason for the license-application rush, however, and it had little to do with the easing of the FCC's rules. Its name was the Great Fuel Shortage of 1973. Its whys and wherefores certainly don't need to be discussed here, but what is important is the impact the fuel shortage had on a small group of men in particular.

The fuel shortage had two immediate effects: It drove the prices of gasoline and diesel fuel sky-high, and it caused the federal government to take a hard look at fuel consumption on the nation's highways. There was very little the feds could do about gasoline and diesel-fuel prices (indeed, government spokesmen expressed a firm desire to keep them high and thus cut down on gasoline and diesel-fuel consumption). They *could* do something about the amounts of fuel used on the highways, though, and do something they did. The national speed limit was dropped ten miles per hour, from 65 mph to 55 mph. This hit the men who drive the big over-the-road trucks right where they lived, and gave them three very difficult problems to overcome. First, fuel prices had rocketed up, while their earning potential hadn't, and the federal government apparently was totally unprepared to do anything about that; second, strict enforcement of the new and much disliked 55-mph speed limit made it nearly impossible for them to make the number of miles that they had to make each day in order to turn a profit; and third, diesel fuel, at least for a time, got downright scarce. The truckers had been into Citizens Band radio for some time, and now they put it to good use. They used it to find fuel when their regular stops had gone dry, and when they violated the speed limit they used it to keep a weather eye out for the dreaded highway patrol. Using Citizens Band radio, these enterprising men created their own radio network. Its sole aim was to keep the participants informed of the whereabouts of highway patrolmen, called Smokies, and to stay informed about which truck stops could be counted on for adequate fuel supplies and which couldn't. Using his radio, a trucker could blast along as fast as his load and the prevailing conditions would let him; it was almost as it had been under the old speed limit. When a call came in over the CB radio pinpointing the location of a prowling patrolman, the trucker obligingly lifted his right foot just a bit until he got clear of any danger of a traffic citation, then went as fast as he could go once again. That's what they're doing to this day, and that's what a lot of other people are doing as well.

A third thing the truckers could do was attempt to demonstrate to the general public just how much of a mess they were in, and just how much that public depended on the goods hauled by the truckers. Passing the word via Citizens Band radio, they organized a general strike.

The strike, of course, got national attention, which was just the idea. For a time, the average truck stop often had as many newsmen and photographers crawling around it as it had truckers. It immediately became apparent to the newsmen that Citizens Band radio had been one of the primary tools used to organize the strike, and word of this and how CB was used in the truckers' everyday routine soon was in every newspaper and news magazine, and was blared out over the airwaves by commercial radio and television. As had happened twice before, the heavy media coverage generated heavy interest. The public was taught about Citizens Band radio for once and for all this time, and it's unlikely they'll ever forget the lessons learned. They learned how truckers used their CB radios to keep traffic cops off their tails, and they figured that if the truckers could do it, so could they. Law-abiding citizens who didn't care a whit about speeding found an interesting tool that they could use to keep in touch with home or with the office, and as a result people began buying CB radios in unheard-of numbers. Sales of CB radios have been so great that some marketing experts consider CB radio the "glamor item" of the year, and most manufacturers of Citizens Band radio sets are selling their products faster than they can make them.

It's like so many other things: The incredible boom in Citizens Band radio is not the result of one small factor. It didn't happen just because of the rules revision, or just because of the truckers and their problems, or because of the glaring eye of the national press, or because of the fuel shortage. All these factors combined to put CB radio on the map, and now that it's there, CB radio has shown no sign whatsoever of slackening.

4

PICKING A CB SET:

On Knowing Where to Start

Getting started in CB really isn't terribly complicated, but to get the best service now and in the years to come, and to get the best radio for your money, a little evaluation and a few simple choices are called for. The choices come in two stages. The first involves deciding how you're going to use a CB set, what you need from it, what you want from it, and what you want to spend. We'll discuss all that in this chapter, along with some additional items of interest.

The second stage is getting down to the specifics of deciding just which features are necessary and learning to use the specification sheets provided by most radio manufacturers to compare the performance of one radio with that of another so that you can pick the one best suited to your needs. We'll talk about that in the next chapter.

One of the first things that will affect your decision on what kind of a CB radio to buy is what kind of a person you are and what you intend using your radio for. This isn't as easy as it sounds, because there seem to be two different schools of CBers, and their needs can be quite different. We'll call the two groups "hobbyists" and "functionalists"; sometimes the line that divides the two groups is a fine one, but it's there, nevertheless.

The functionalist is someone with a purpose. He might be a man in a tractor, out plowing ground, who wants to be able to pick up a microphone mounted in his tractor cab or in his pickup and tell his wife he's on his way in for lunch; or maybe he wants to be able to tell a helper that he's out of gas or had a breakdown. He might be a man with a small delivery service, to whom radio communication means fewer unnecessary trips and greater efficiency in general. A functionalist might be a trucker who uses CB to keep track of fuel supplies and traffic cops, or a boatman who wants effective but inexpensive communications with other boats and with the shore.

The functionalist is the person with a specific purpose in mind that he knows CB radio can fill. He's going to want to consider the total cash outlay involved, and he probably isn't going to want one more extra feature on his radio than he has to have to get the job done.

The hobbyist, more often than not, is someone who is simply taken with the magic of two-way radio. He enjoys talking with other CBers; he has a good time and makes friends over the air. He might know a little about the technical side of CB, and for a hobbyist's reasons, it's important that his set perform as close to the maximum legal range as possible. For him, two-way radio is probably more a form of relaxation than a necessity.

The hobbyist might keep an ear open for smokey (traffic cops) reports when he's out on the highway, or perhaps communicate with someone at home, and the functionalist might use his set in the evenings to find out what's going on in the local CB community. In other words, you can be a little of each, functionalist and hobbyist. The way you intend using your set will more than likely have some bearing on what sort of set you buy.

For instance, do you want a mobile unit, a base station, or a walkie-talkie? Those are the three main varieties of CB radio. A mobile unit lives in your vehicle and in most cases is designed and styled to be mounted under a dashboard. A base station is built to stay in one place, usually in an office or home. Some are built in hard, functional metal cases; others are styled with chrome and walnut. A walkie-talkie is just that, a hand-held portable radio with its own built-in antenna. There are advantages and disadvantages to each variety.

A mobile unit has the obvious advantage of keeping you in touch whenever you're out in your car, since wherever your car goes, it goes. Mobile units are generally less expensive than comparable base stations, and are far and away the biggest sellers. The disadvantages are that a mobile unit is inconvenient

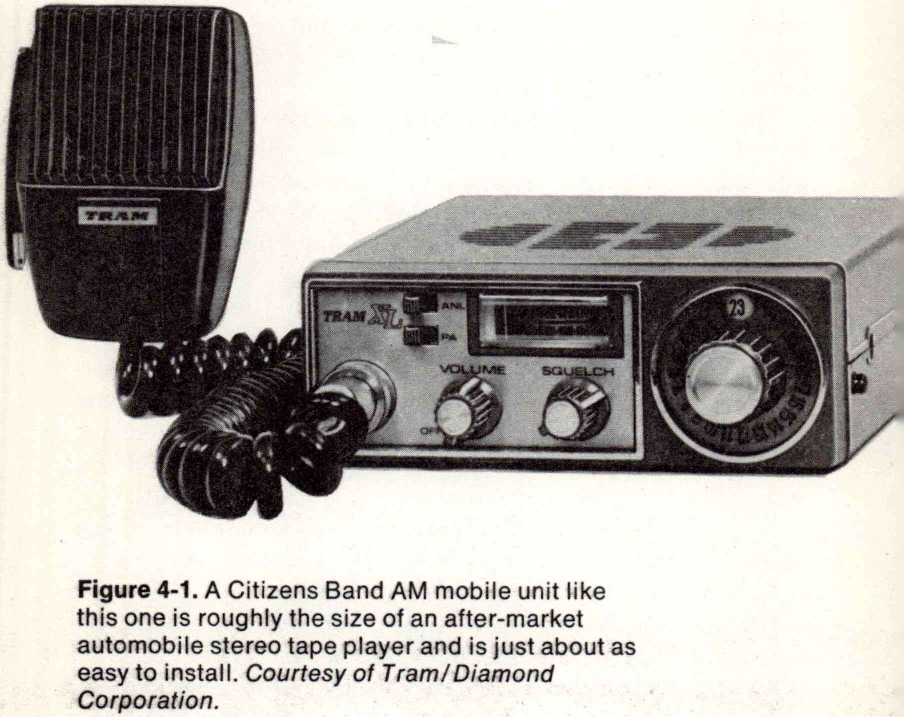

Figure 4-1. A Citizens Band AM mobile unit like this one is roughly the size of an after-market automobile stereo tape player and is just about as easy to install. *Courtesy of Tram/Diamond Corporation.*

to use once the car in which it is mounted has been garaged, and that a mobile unit's range generally cannot match the range of a base station. This has nothing to do with a radio's performance; rather, it is a reflection of the fact that mobile antennas generally do not provide the range that base-station antennas can. Additionally, mobile units pick up more noise than do base stations, most of it coming from your car's ignition and electrical systems.

Which is not to say that base stations don't also have drawbacks. First of all, most base stations are generally more expensive than mobile units with similar features. Second, buying a base station means that whenever you're away from home you'll be without your radio—unless you have a mobile home with an antenna on it and a place inside to set your radio, and most people don't.

There is one major advantage to owning a base station and its companion antenna setup, however, and that's simply that a base station will provide much cleaner sound and considerably better range than a mobile unit can. The reason a base station's sound is generally cleaner than a mobile unit's is that CB radios pick up car noise. And not just noise from the cars in which they're mounted; the noise from passing cars and

Figure 4-2. Base stations come in all descriptions. This one is functional in appearance and comes in a metal case. *Courtesy of Teaberry Electronics Corporation.*

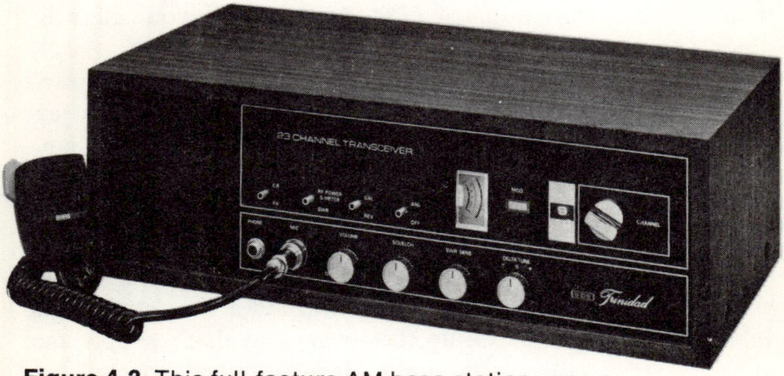

Figure 4-3. This full-feature AM base station uses a wood cabinet and bright metal to look like a fine piece of stereo equipment. *Courtesy of SBE Linear Systems, Inc.*

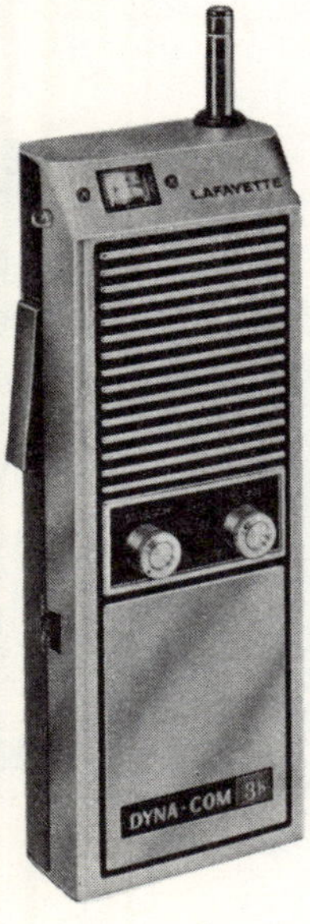

Figure 4-4. If its relatively short range doesn't interfere with your needs, a handheld Citizens Band radio might be your best bet. On this model as on most others, the antenna telescopes up out of the top of the radio's case. *Courtesy of Lafayette Radio Electronics Corporation.*

trucks can also be picked up. It comes from distributors, spark-plug wires, and alternators, and comes across the air as popping and buzzing sounds that rise and fall in tempo and pitch as the speed of the offending engine rises and falls. When the noise is coming from your car, most of it can be deleted; when it's coming from another vehicle, there's not much that can be done. Naturally, a radio installed in a house or office is isolated from these noise sources to a much greater degree than a mobile unit.

The range of a base station is wider than that of a mobile unit because base stations usually have much better antennas. Practical considerations limit the length of mobile antennas, and thus the radio's range. Mobile antennas have to be able to duck into garages, and they can't be so long that they knock out the overhead light fixtures at your local service station. For the best overall performance, the only answer is to drill a hole in your car's roof and mount the antenna dead center. Many people shy away from this hole-drilling, however, and choose instead a compromise mounting position that does not mar the car's sheet metal, but may not provide the performance a roof-mounted antenna would.

Unlike the antenna for a mobile unit, the antenna for a base station has no height limitations dictated by practicality. The limitation is imposed by the FCC. It allows 20 feet above a structure or a total of 60 feet on a tower. If you mount your antenna 20 feet above the peak of your home's roof, and that peak is, say, 14 to 16 feet above the ground, the total antenna height is approximately 35 feet, a considerable improvement over the height of any mobile antenna. The result is much greater range.

A walkie-talkie? Neat little units, especially the 5-watt, 23-channel subspecies, but, like everything else, they too have limitations. The most important is range. If you're never going to be more than a few miles from another radio, you'll be in good shape, but if you get more than just a few miles away, you'll be out of luck, because a walkie-talkie's short antenna won't be able to do the job. Most walkie-talkies can be hooked up to your mobile unit's antenna, but that ties the radio to your car, where it probably won't perform as well as your mobile unit. The second disadvantage involves the short life span of the power supply. A battery-supplied 5 watts is the input (in some it's 3.5 watts and in others still less), and if you use the unit much, you're going to be replacing a lot of batteries. If you're a member of a surveying team, though, or if you're involved in cross-country skiing or on-foot search and rescue, a walkie-talkie might be just the thing.

By the way, don't confuse CB walkie-talkies with the kind the kids get for Christmas and birthdays. Those work on CB, often on channel 14, and require no license. But it's illegal to use one to talk to a licensed station, and the sets won't take much everyday use and abuse before they're ready for the trash collector.

Next decision: Do you want a set that uses tubes, or do you want one that depends on transistors? That's something the various manufacturers of Citizens Band radios have practically decided for you; there just aren't too many tube-type sets being built these days. But there are a few, and in most cases the unit involved is a base station. Tube-type stations have two drawbacks: 1) Tube units draw more power than do transistor units, and 2) they take up more room. This is much more critical with a mobile unit than it is with a base station, and if you figure you need a tube unit, just shop around a bit; there are some excellent radios available.

Most of the CB radios you'll be seeing as you shop will come in one piece, but there are some alternatives. They exist at least partly because a radio is made up of three parts: a transmitter, a receiver, and a power supply. These pieces can be mounted in separate boxes and strung together by electrical cables, on separate chassis that are all housed in a single box, or all mounted in one small box, with certain circuits integrated so that they can be used for both the sending and receiving functions. A unit built like this is called a transceiver and is far and away the most commonly seen type of set. Because certain circuits are of a dual-function nature, certain other circuits have been eliminated, and that means the transceiver is at once more compact and generally less expensive than its nonintegrated counterparts.

There are other alternatives, too. Being seen more and more these days is an item called a CB converter. It's a small electronic box that you hook up to the radio in your car and then screw to your car's dashboard. When you turn it on it converts your car's radio to a CB receiver—and only a receiver. A converter will not enable you to send CB signals. But if your primary interest is listening for the location of traffic cops as relayed by truck drivers, a converter might be just the thing. Tuning one is a little tricky, because you use the radio's tun-

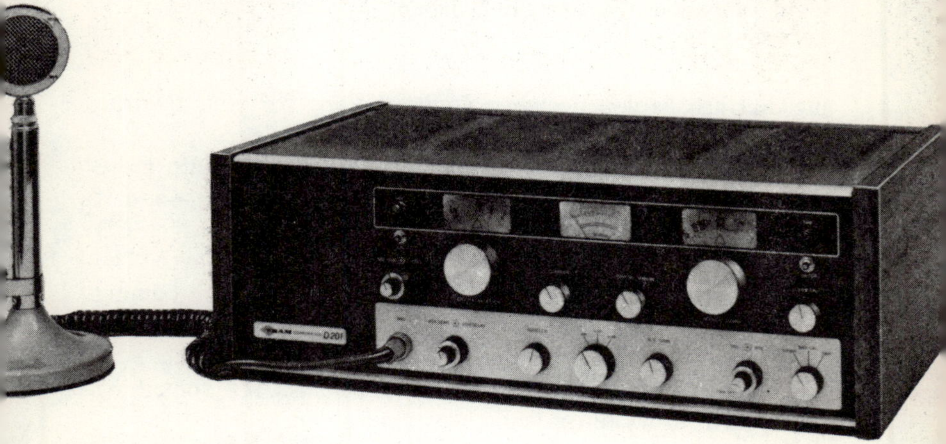

Figure 4-5. Most base stations sold these days are solid-state, but some do use tubes, as does this model. Tube-type units generally are a little more bulky than transistor units, and generally consume somewhat more electrical power. *Courtesy of Tram/Diamond Corporation.*

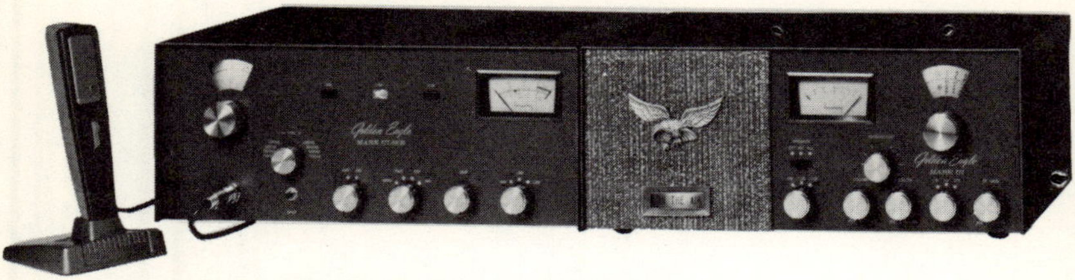
Figure 4-6. Not all Citizens Band radios are transceivers. This one, for instance, comes in two pieces, a receiver and a transmitter. It's a beautiful unit, but considerably more bulky than a solid-state transceiver. *Courtesy of Browning Laboratories, Inc.*

ing dial just as you do when you're looking for your favorite station, but you'll get the hang of it after a while.

Another interesting device that is finding its way into more and more mobile applications is the combination set—and it can be a combination of a number of things, including CB and AM/FM, or CB and AM/FM and stereo tape player. The people who make these kinds of units say they're the coming thing and that they work very well; the people who don't make them say they present too many compromises, that their very compactness can lead to circuitry problems. It's something you'll have to make up your own mind about.

Two more possibilities are the unit with fewer than twenty-three channels and the kit. If a functional, low-cost radio is what you're interested in, then a unit with less than the full complement of channels may be the best bet for you, and there are several about. It's the kind of thing that might be a particularly good buy if you're interested only in talking to truckers. And if the one you have in mind doesn't have the channels in it you need, new, different crystals can be plugged into the set to provide your choice of channels. One point of view is that with twenty-three channels available, and with the CB airwaves as crowded in some areas as they are, it's foolish to buy anything that is going to restrict you to less than full use of all the CB radio airwaves. Is that a valid argument? That's up to you, your wallet, and the way you use your set.

If you're handy with tools and have a little basic knowledge of electronics (or want to get some), a kit may be just the thing for you. A kit could save you a bit of money, and if you're like most people, you'll get more than a little satisfaction out of putting the radio together. The job isn't all that hard because the unit that contains the crystals comes pretuned, prewired, and sealed so that it will meet the government's rather exacting type-acceptance requirements for CB radios. Type acceptance, by the way, means merely that before any CB radio can be sold it must meet certain minimum technical requirements. This is how the FCC insures that CB units conform to the letter of the law in areas like use of proper frequencies and maximum power output.

So when you buy a kit, you can count on the hard part being done for you. All you have to do is stick the light stuff together. Just one thing: Kits do take time to assemble, so keep that in mind while you're making your decision.

There is one more possibility that has not yet been covered, and it involves using a mobile unit as a base station. To do this you have to buy a power supply that converts the mobile unit's appetite for 12 volts DC to an appetite for 110 volts AC. The advantage is that you can use your mobile unit both in your car and at home; the disadvantage is that most mobile units are designed to be mounted on a dashboard. If you're going to take one inside for use on a power supply converter, you have to unmount and completely disconnect it, which can be inconvenient and time-consuming.

So now you've made your choice: mobile, base, walkie-

Figure 4-7. If you're on a budget, a CB radio with less than the full twenty-three channel range might be the answer. This one has five channels; some have fewer, some have more. *Courtesy of SBE Linear Systems, Inc.*

talkie, converter, kit, or alternate power supply. If you've thought about it at all, you probably have some ideas about which will suit you best. What you should also know, then, is that no matter what you decide will be best for you, the innards of these various kinds of sets will all give you similar performance—they all have to meet the same set of government standards, remember? What remains is to choose a unit on the basis of what additional features the unit has, and on the basis of cosmetics—which set has, in your eyes, the prettiest face.

Pretty faces usually are a matter of personal preference, and in terms of making up your mind about what you do and don't like in that regard, this book isn't going to be much help.

What you do have to decide, though, is which features are worth having and which are not. To make up your mind about which ones you want and which ones you don't, you'd best start by knowing which features are considered basic, non-optional, must-have items. Four features fall into this category: They are the on/off/volume switch, the squelch knob, the channel selector, and the push-to-talk switch.

The on/off/volume switch is just that. It's an on/off switch that is built into, or ganged with, a device called a potentiometer that works more or less like a rheostat and can be turned clockwise to increase the volume the set's speaker is producing, or counterclockwise to decrease the volume. So when you receive a signal that is too loud, you can turn it down, and when you get one that isn't loud enough, you can turn it up.

The second basic is a squelch knob, which also is turned clockwise when you wish to increase the amount of work it is doing. The work that it does is simply to cut down on the amount of noise coming in over the set's speaker while the receiver is not in action picking up a signal. It works like this: No matter what kind of set you choose and no matter where it is going to live, your antenna is going to pick up a certain amount of background noise. The noise comes from various sources, and they aren't important right now. What is important is that you have to get rid of the noise between signals, because it will drive you up the wall if you have to listen to it for any amount of time. Hence the squelch knob. It establishes a sort of electronic threshold that a signal must overcome if it is to be heard. It cuts down on background noise, very weak signals, static, and distant signals brought in by skip. You can use a radio with the squelch turned all the way down, and since a strong signal will override background noise, you'll receive just fine. But when the signal is gone, you'll get static and a constant hissing sound. With the squelch knob properly advanced, your radio's speaker will remain completely silent unless there is a bona fide signal coming over the air.

The third basic control is the channel selector. It's usually a round dial with the numbers from one to twenty-three imprinted on it, although some newer sets rely on a dial and numbers shown by electronic digital display. Regardless of type, the function is the same and is straightforward: You use it to choose the channel you want. It works just like a TV channel selector, with channel detents that line up with each numbered channel. You decide which channel you want and then turn the knob to display that channel.

Last, there is the push-to-talk switch. When keyed, it activates the transmitter and opens the mike, and when unkeyed, its usual position, it keeps the receiver on and in search of a signal.

These are the basics, and you can take them for granted if you wish. But there are a lot of other common features that you can't take for granted; you'll have to make up your mind about each of them. Ultimately, the majority of them will not increase your radio's performance significantly, but one or more of them might make a radio a little easier to use and more fun as well. The purpose of this chapter is not to tell you which of them are good and which are bad, which you need and which you don't; you'll have to decide that for yourself. It will, however, attempt to give you a working understanding of what each control does, and why, so you can make a reasonable decision about what you need and what you don't.

One of the most commonly found "extras" on mobile sets is a switch, usually a small sliding switch, labeled CB/PA. In the standard operating mode the switch would be left in the CB position and the set would operate as a radio. But with the switch in PA mode, and with a small external speaker set up on a roof rack or under your car's hood, the set will function as a public-address system.

An automatic noise limiter usually has a sliding switch to activate it. It's labeled ANL, and when the switch is activated, the limiter works to mitigate atmospheric static and general noise, as well as some internal radio noise.

A noise blanker, also usually controlled by a sliding switch, does the same sort of thing the automatic noise limiter does, but works only on certain kinds of noise; it's especially effective against noise made by car ignition and electrical systems.

A feature many people consider largely unimportant is a transmit light. It lights up when the transmitter section of your transceiver is activated by the push-to-talk button.

Generally considered to be much more of a necessity than a transmit light is an S meter. Like many of the features discussed in this chapter, an S meter has no bearing at all on how well a radio performs; it does, however, allow you to read the strength of incoming signals and can be rather handy. Most S meters are marked off in units from one to nine, and in decibels by tens above that. The units between one and nine, by the way, are referred to in CB slang as "pounds." So if you hear someone ask for a radio check, and then you hear someone else come back with "You're hitting me with seven pounds," you know that the second operator has monitored the first CBer's signal on his S meter, and that the meter's needle went to seven during the signal.

At least one manufacturer has replaced the S meter on one of its units with a series of light-emitting diodes. In this set, instead of a needle going across a dial face, the intensity of the light increases as the strength of the incoming signal increases. Some people think this sort of system is easier to see. You'll have to make up your own mind.

While we're talking about built-in meters, let us not forget the built-in SWR meter, something you normally find only on certain base-station units. SWR means standing wave ratio; it will be discussed in another chapter. For now, know that a built-in SWR meter is a neat gadget, but if you also are planning on owning a mobile unit, you'll probably be better off with a separate SWR meter that isn't tied to the circuitry of your base station.

A mike-gain control is a handy feature, but one most of us are going to have to do without, because the FCC is no longer type-accepting transceivers with built-in mike-gain controls. What the control does is vary the amount of voice power, or modulation, reaching the transmitter. By watching your S meter when you transmit and adjusting the mike-gain knob, it is possible to speak into the microphone at any level of voice that seems appropriate and to control whatever distortion might occur.

A feature many experienced CBers particularly favor is an RF-gain control. RF stands for radio frequency, and the RF-gain control affects the radio frequencies that the antenna picks up and works them over before your receiver picks them up and translates them into audio frequencies for your speaker to spew out. An RF-gain control helps keep a really strong signal (such as might come from a CBer very close to you, or from one who is not so close if skip is in) from overloading your receiver and bleeding onto other channels. Using an RF knob, you have control over just what portion of a strong signal your receiver picks up. Ideally, you would turn the signal down as far as you could with the RF-gain control, and use the radio's volume knob to bring the sound up to a listenable level. In other words, an RF-gain control can be used to give your set more selectivity and less sensitivity. This is the sort of thing a CBer might find especially useful if he lived in a heavily populated area, and it's the sort of thing he could rightfully expect to find on the more expensive base stations.

Next is delta-tune, and no, it won't bring you Cajun music from the Mississippi River delta area. All CB sets these days are crystal controlled and tuned to tightly fixed frequencies, but because of production tolerances and whatnot, some sets can be off their assigned frequencies by ever so little. When you run across a set that is out of tune, you'll find that its signal is very difficult for your receiver to translate, and that's where delta-tune comes in. Most delta-tune knobs have three positions: a median position, a plus (or over) position, and a minus (or under) position. Ordinarily the switch is left in the median position until the need to tune your receiver to an out-of-tune transmitter becomes evident. Then you switch your delta-tune from median to plus or minus, thus adjusting your receiver slightly one way or another, until you find out which way results in the best incoming signal.

Channel 9 has been designated by the FCC as Citizens Band radio's emergency channel, and there are many people who want to be able to monitor channel 9. If you're one of them, and if you want to keep your radio tuned to your area's hailing (or call) channel for use in regular, nonemergency communications, a radio with a built-in channel-9 monitor might be just the thing for you. On some sets the monitor is a light that goes on when there's activity on 9; on others it's a system that lets 9 break into whatever channel you're on. Both methods are effective; you'll have to decide for yourself if you want either system, and if so, which one you would prefer.

Many experienced CBers wouldn't want to be without the capability of handling an auxiliary speaker, and industry figures indicate that the average CBer will probably buy two auxiliary speakers during the lifetime of a given mobile unit. An auxiliary speaker is one that is mounted in a separate baffle box. Both the speaker and its box are designed to handle voice frequencies and provide near-exact reproduction, and chances are that an accessory speaker will enable you to hear more things more clearly because of this. In order to use one, your radio must have a jack into which you can plug the speaker. Before you decide that you really need an auxiliary speaker, find someone who has one and listen to it a bit. Then make up your own mind about whether or not it's really necessary.

Clocks, tone controls, and AC/DC switching are features that normally will be found only on base stations. Do you need a clock on your base station? Well, if you don't wear a watch, and if your radio is located in a position from which you can't see a clock, perhaps you do.

A tone control on a CB radio works just the way a tone control on a stereo set does; it adjusts the timbre of the audio signal delivered by the set's speaker. Some people consider it a pleasant addition, while others consider it largely unnecessary. AC/DC switching, however, is another matter. Suppose you have a base station with AC/DC capability and you have everything set up to utilize that capability. (It takes a 12-volt car battery and a few wires, by the way.) Suppose the power lines that feed your block, or your farm, or your homestead away out in the desert, are knocked down or blown over. Your unit is dead, useless, right? Wrong. You've got it hooked up to an alternate power supply, remember? All you have to do is turn the radio on, call for someone who still has power, and ask him to report your problem on his telephone.

Two other features found on some radios are of particular interest. They are a detachable microphone and remote channel switching. A detachable mike is not a feature usually found on price-leader sets for the simple reason that it costs more to build it. Such a convenience, however, gives you the option of replacing the mike with a specialty mike from the accessory market, should you ever feel the need.

Remote channel switching is mentioned here because the

control for this is usually mounted right on the mike. It too is a handy feature, allowing you to mount your mobile CB set wherever you want without having to lean all the way over to the set to switch channels. You just press a button on the mike to do that. A convenient feature, but one you'll have to pay to get. Is it worth what you'll pay for it? As with all the rest of the features covered in this chapter, that is something you will have to look into and decide for yourself.

5

CHOOSING THE RIGHT SET:

Separating the Wheat from the Chaff

So now you've decided what sort of Citizens Band set will best fill your needs, you know which features you'd like to have on it, and you know approximately how much you want to spend on radio and antenna. The next step is deciding exactly which set to buy, and from whom.

Finding a store that sells CB sets shouldn't present much difficulty. Large department stores are selling CB sets these days, and so are mail-order stores, discount stores, drugstores, and supermarkets. And don't forget that you can also buy CB sets at stores specializing in CB radio. Which one is your best bet?

Chances are that it will be to your advantage to buy your set from a store that specializes in CB radio, because most CB shops can offer you considerable variety, have their own service departments, and can and will do installations. Department and mail-order stores may offer more shopping convenience, and a discount store may offer a lower price on a given CB unit, but many such stores offer just one brand, and that makes comparison shopping more difficult.

Also, if your radio should need to visit a service department you could run into trouble, because most stores are interested only in selling CB radios, not in maintaining a shop to service them after they've been sold. Simply put, their job is to move the product, not to service it.

Discount, mail-order, and department stores usually do not offer installation either. Installation isn't difficult, and a little farther on in this book you'll find step-by-step instructions and photos so you can do it yourself. But maybe you're just not a do-it-yourselfer, and if not, you'd best buy your set from a dealer who will install it for you.

Another problem with buying a CB radio from a discount or department store is that while the salespeople might be ready and willing to answer questions, chances are that they will not be able to answer them as satisfactorily as someone who runs a CB shop. So figure on doing some shopping in your local CB store.

But first you have to find one. Look in the yellow pages of your telephone directory under "Radio Communication Equipment and Systems." If there's just one CB store in your locale, aim for it, but if there's more than one, you might consider starting your search by going into the stores that offer you the widest selection of equipment. Give yourself a chance to chat with the salespeople and find out just how willing to help and to answer questions they are.

Just for the sake of discussion, let's say that you've decided a base station will best fill your needs, and that you want your new radio to be equipped with several features that usually are not considered standard. You're going to do some serious comparison shopping and you're going to look into every properly equipped base station that you can find, so you'll need some way to keep track of the notes that you'll be taking. Get yourself a steno tablet and list on the first page the features that you consider must-haves and those that you consider optional. On succeeding pages you can list the various radio sets you look at, allowing a page for each set. List the name, model, price, warranty, and performance specifications. This will put a reference guide in your hand, one you can use to make direct, head-to-head radio comparisions. Carry your tablet from store to store and note the strong and weak points of each set you see. When you're finished shopping, you can go home, pour a cup of coffee, and take as much time as you need to decide just which of the radios you've seen is the one you want. This all may take a little time and trouble, but it beats going into the first store you see, slapping down two hundred and fifty dollars for the nearest radio, and finding out later that you could have done much, much better for your money.

Where to start? Walk on into a CB store, tablet in hand, and tell the man behind the counter what sort of set you want and which features you consider necessary. Ask him to show you any radios he may have that offer the features you require. Remember, you're going to list each radio on a separate page, with its name, model, features, and price. So start by finding out what the term of the manufacturer's warranty is, and whether or not you can obtain a service manual for the radio. You want a year's warranty because that's standard, and you want a service manual because chances are good that you'll own your CB set for quite some time. Most CB radios are sturdy, and there isn't too much that can go wrong with them. But radio makers do change their model lines from time to time, and it could be that somewhere on down the line the radio you're about to buy will be phased out of production. Or it might be that you'll have to have your radio fixed sometime in the distant future by a shop that doesn't handle your brand, has never seen a radio like yours, and does not have the circuit diagrams. That's when you'll whip out the service manual that you bought when the radio was new, and that's when you'll be glad you have it.

The basic specifications of most CB radios are roughly the same: 5 watts input, a maximum of 4 watts output, twenty-three channels. But there are many other specifications that are not the same, and that's where things begin to get a little technical. Each radio manufacturer publishes the exact specifications of each radio he builds, however, and with a little preparation you can look those specs over and see which radios are the performers and which are not.

You have to start some place, and it might as well be with the radio's circuitry. Most radios these days are superheterodyne circuitry, which is a great improvement on an older type called superregenerative. Superhet, as it's sometimes called in radio slang, provides more complex circuits, but offers greatly improved sensitivity and selectivity. Superhet comes in either single conversion or dual conversion; these terms describe the way a radio's receiver section handles incoming radio signals. A single-conversion radio gives you a receiver with a low noise level, but its adjacent channel selectivity, its ability to keep signals from other nearby channels from bleeding onto the channel you're using, usually isn't as good as that of a double-conversion radio. Double-conversion receivers put incoming signals through a couple of additional steps before they

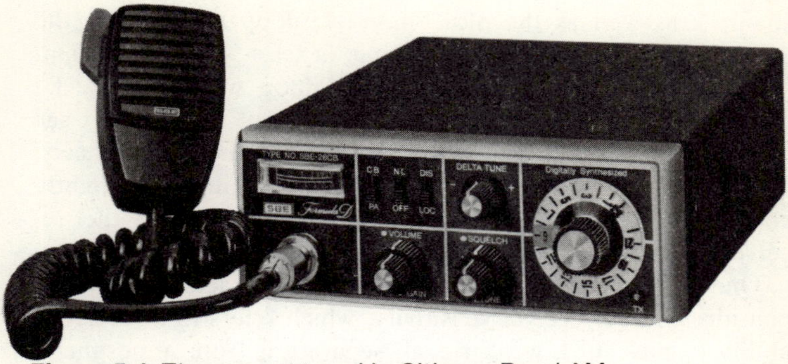

Figure 5-1. The newest trend in Citizens Band AM radios is phase-lock-loop, a process that synthesizes all twenty-three channel frequencies by utilizing one to three crystals. *Courtesy of SBE Linear Systems, Inc.*

are sent to the radio's audio portion. The result is a receiver that gives greatly improved adjacent-channel selectivity, but produces slightly more noise than a single-conversion set. So it's a trade-off, and if you can, you'd probably be well advised to opt for the improved adjacent-channel selectivity of a double-conversion set.

Another thing you should consider is whether or not you want a radio that uses the phase-lock-loop process. This is a relatively new system that involves a revision of the way a CB radio signal is generated. Most CB radios use from twelve to fifteen crystals to synthesize each of the twenty-three different radio frequencies available. Usually one crystal is used both to transmit and receive, another is used just in send, and another is used just in receive. These crystals all vibrate at given frequencies, and the frequencies are mixed to provide the correct CB channel wavelength. When you switch channels, you switch sets of crystals. Phase-lock-loop changes all this, and a radio built around this method can use a total of from one to three crystals instead of fourteen or fifteen. By means of some complex new circuitry, these few crystals can be tuned to deliver as many different frequencies as called for.

In general, phase-lock-loop radios are more precisely on-frequency when they send, and they're a bit more expensive than the usual. If the FCC decides to provide new frequencies for CB radio, depending on where those frequencies are located in the radio spectrum it is possible that a phase-lock-loop radio could be modified so that it would work on the new channels as well as the old. Phase-lock-loop is definitely the coming thing. But for the time being, at least, you'll pay for it.

Heterodyne circuits and phase-lock-loop will probably be mentioned on manufacturers' spec sheets, but if you don't see them, ask. What will surely be mentioned on those spec sheets, though, are receiver specifications. CB transmitters are tightly controlled by the FCC's type-acceptance rules, but the receivers are not. So look hard at these figures; they're important.

One of the most important receiver characteristics you'll consider will be receiver sensitivity, which refers to a receiver's ability to pick up a weak signal, separate it from atmospheric and electrical noises, and amplify it so you can understand it. Sensitivity figures are recorded in microvolts, and the smaller the microvolt number, the more sensitive the receiver. A receiver's sensitivity is usually stated as signal-plus-noise over noise, just like a math equation, and the smaller the number is, the better. It might be stated as 10 decibels at .5 microvolts, or 10 decibels at .3 microvolts. A well-designed receiver should provide a minimum sensitivity of 1.0 microvolt.

Adjacent-channel rejection is next in importance and is proportional to sensitivity. It refers to the ability of a receiver to keep signals being sent on the next adjacent channel from slopping over and bleeding through onto the channel you're tuned to. The more rejection the better; look for a minimum of 40 decibels, with 45 decibels better and 50 decibels still better.

Spurious rejection is similar to adjacent-channel rejection, and has to do with a receiver's ability to lock onto a signal, much as an FM receiver can be locked onto an FM signal. It helps keep stray radio signals from muddying up your channel, and the minimum acceptable number is 40 decibels. Higher decibel numbers mean better rejection.

You'll also want to consider the radio's audio power output. This is the amount of power your receiver sends to your set's speaker. If you plan on buying a radio that has a PA function and you plan on using the PA, or if you're going to mount the radio in an area with a good deal of environmental noise, a high audio-power-output figure will be important to you. Look for a receiver that provides audio output of 5 or 6 watts. If you do not want a PA function and environmental noise is not a consideration, all you'll need is enough power to drive the radio's speaker, and 3 or 4 watts will do that just fine. Don't go below 3 watts, though.

So far we've talked only about the receiver. There are some important specifications that apply to the transmitter, and you should know about those, too. Most important is power output. According to the FCC, the maximum is 4 watts, and that's what you should look for. Some radios produce 3 watts; some produce more, some less. The more power your set produces, the cleaner its signal will be and the greater its range. So you want a set that has the maximum legal power output.

You'll also want to look into the percentage of voice modulation. This has to do with the superimposing of voice frequencies over radio carrier frequencies. Modulation means mixing the two, and the percentage of modulation is the amount of voice signal compared to the amount of carrier signal. An equal mixture is ideal and is listed as 100 percent modulation. The FCC requires that CB sets contain modulation limiters to keep you from overmixing voice frequencies, so many manufacturers list modulation as limited to 95 percent. Anything less than 100 percent means that there's just that much of your signal that isn't getting out over the airwaves.

Some of what we've discussed may sound pretty complicated, but really it's not. You're simply looking for a few easy figures to keep in mind when you go radio shopping. You want a warranty of a year; you want a service manual; you want a set of the double-conversion type; you may wish to consider phase-lock-loop; you want receiver sensitivity of 1.0 microvolt or less; you want adjacent-channel rejection of at least 40 decibels; you want spurious rejection of at least 40 decibels; you want audio output of at least 3 watts; you want a full 4 watts of power output; and you want as close to 100 percent modulation as possible. The only thing that is even slightly complicated is digging the information you need out of spec sheets or out of the CB salesman. If the information you want isn't available for a specific radio or if the information is stated in a way that neither you nor the salesman can understand, pass that radio by. It might be a fine set, but your objective is to come away with the very best radio you can get for your money, and the only way you can do that is to insist upon getting adequate, comprehensible information about it.

6

SINGLE SIDEBAND:

What It Is, What It Does

If you're a lover of things exotic (cars, food, companions), you'll be delighted to learn that a singular piece of exotica exists in the world of Citizens Band radio. It's special enough to deserve its own short chapter in this book, and its name is "single sideband." It's usually referred to as SSB or just sideband, and if you're going to get into it, your bank account also needs to be just a little bit exotic. These sets are expensive, with typical mobile units selling for approximately three hundred and fifty dollars and typical base stations selling for approximately four hundred dollars. So sideband sets cost quite a bit more than regular AM sets, but they deliver more, too.

AM sets, those that deliver straight, no-nonsense CB radio, are called that because they use a method known as amplitude modulation (AM) to add the voice signal to the radio's carrier frequency in the transmitter. The carrier, by the way, is the channel-operating frequency, and is a healthy band of electromagnetic energy. When you send on an AM set, your voice is amplified and combined with a carrier to modulate it. You send a carrier and a pair of sidebands, which are smaller mirror images of energy that run along each side of the carrier. These sidebands carry the voice frequencies that a receiver translates back into audio, and the carrier makes sure that everything stays on frequency.

So an AM transmitter has got a signal to put out, and that signal is composed of three parts. Now, you already know that CB radio is limited to a maximum power output of 4 watts, so what you have is a relatively small amount of power pushing a three-part signal. Single sideband changes this balance of power more than a little. A single-sideband transceiver adds the voice to the radio signal in such a way that the carrier and one of the two sidebands are unneeded and can be suppressed before transmission. All the power is put behind the remaining sideband, which effectively gets three times the power it got before it was isolated. You are now using 4 watts to push one-third of the signal an AM station produces. Sideband transmitter power ratings reflect this change. Most of them offer, instead of an AM set's 4 watts, 12 watts peak envelope power (PEP). It amounts to a far more efficient use of energy and of the radio airwaves.

The chief disadvantage of sideband is its price, which we discussed earlier. Some CBers consider the mode's inability to communicate with a straight AM set a disadvantage, but those with both base and mobile sideband sets consider SSB an advantage, since AM stations rarely can understand a sideband signal and this results in a bit more privacy than one gets on AM. AM radios can hear sideband signals, by the way, because both broadcast on the same frequency, but a sideband signal is rarely understandable to an AM set.

The chief advantages of sideband are the extra range it provides, the fact that its radio signals are a good deal more clear and free from interference than those of an AM set, and the fact that sideband gives you a choice of sixty-nine operating channels. Sideband sets are not affected by atmospheric interference as much as AM sets are, and their signals are more easily understood on crowded channels.

That huge number of operating channels isn't magic. It works like this: A sideband set is still a CB radio, and it still has the twenty-three-channel AM capability. It also has a couple of extra switches, and both of them are of primary importance. The first has three modes, and they're labeled AM, USB (upper sideband), and LSB (lower sideband). With the switch in the AM position you have access to CB's usual twenty-three channels. With it in the USB position you have access to the twenty-three upper sidebands, and with it in the LSB position you have access to the twenty-three lower sidebands. Adding them all up makes a total of sixty-nine, but to say that a single-sideband set has sixty-nine channels actually is not completely accurate. A better way of putting it would be to say that a sideband set has the ability of using each of the twenty-three channels in three different ways, since a station using 23 AM, a station using 23 USB, and a station using 23 LSB are all using channel 23, and a receiver set to channel 23 will hear all that is being sent. It's not likely that three stations as described here could really use the same channel at the same time, but one sidebander could use upper 23 while another used lower 23.

With sideband taking up less than half the space in the airwaves that an AM signal does, and with the CB radio bands becoming more and more crowded, many observers feel that sideband is the coming thing. If you've decided that a sideband set is what you need, look for the same things in a sideband radio that you would look for in an AM transceiver, because they all function the same way. With one exception.

You recall that a couple of extra switches were mentioned, and that one of them was the AM/USB/LSB switch. The other one operates a thing called a clarifier; it's an absolute necessity on a sideband set, and all of them have one. It works more or less like a delta-tune control, in that it helps make unintelligible signals more intelligible. When a signal is produced by one sideband set and received by another, the receiver portion of the receiving transceiver inserts a carrier back into the incoming sideband to permit demodulation, separation of the voice and radio frequencies. But it needs some help, and that's where the clarifier comes in.

When an SSB signal is first intercepted, it often is nearly impossible to understand, and can sound completely distorted. It might come across hollow and boomy, as though someone were talking to you through a 20-foot-long piece of 4-inch steel pipe, or it might come across as high, squeaky noises that sound a good deal like Donald Duck at his angry best. The clarifier is used to ungarble this signal so that you can understand it. It generally has to be used whenever you intercept a new station.

How do you know if sideband is for you? As with everything else presented here, that's something you're going to have to decide for yourself. If you live in a large city where the airwaves are very crowded, you might want to give sideband some strong consideration. But if you live in the countryside or in a small community, you may be one of just a very few

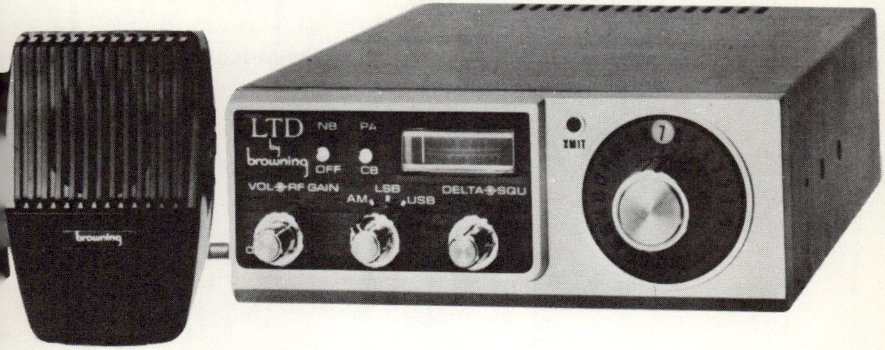

Figure 6-1. A single sideband mobile unit looks much like an AM unit. The giveaway is that a sideband unit will have the AM USB/LSB switch and a clarifier control. *Courtesy of Browning Laboratories, Inc.*

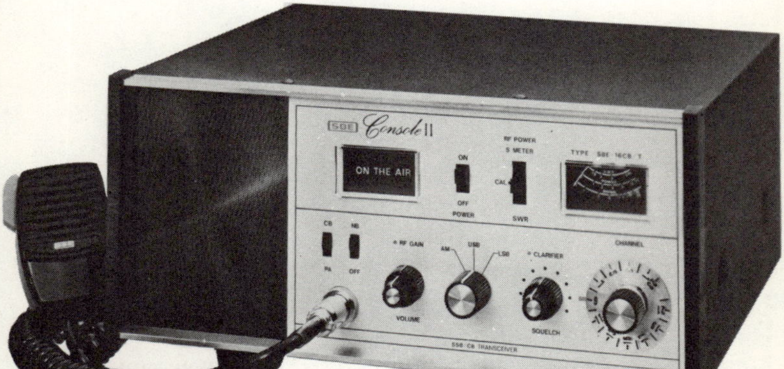

Figure 6-2. This SBE Console II is a single-sideband base station with a built-in combined SWR and power meter and an RF-gain control. It also has a clarifier control, which is something all sideband units must have. *Courtesy of SBE Linear Systems, Inc.*

who are into sideband—and it's no good at all unless you have other people to talk to. If you want the ultimate in CB communications between a base and a mobile, you might want sideband, but if you don't, you won't. So weigh everything carefully, and ask yourself just how badly you need to get involved with this most sophisticated form of Citizens Band radio. Then check your bank account and the price tags on sideband radios, and ask yourself again.

7

ALL ABOUT ANTENNAS:

Getting the Best of a Black Art

There seems to be a tendency on the part of many people, when they buy their first Citizens Band radio set, to put a considerable amount of time and energy into finding just the right radio with just the right features at just the right price. That done, they turn around and buy the first antenna that happens to fall on them from the store shelf.

The willingness on the part of many CBers to get this aspect of their business over in a hurry can be forgiven, because if anything ever possessed most of the proper qualifications for being labeled a black art, it would have to be the science of antennas. It's a highly technical science, and one that has some pretty complicated secrets, but these secrets can be made to reveal all the most important facets of Citizens Band radio performance. All it takes is a little time.

That you should spend the time and thought to properly understand antennas is not even arguable, because of one irrefutable fact: The most expensive CB radio in the world, hooked up to a cheap, low-performance antenna, absolutely will not perform as well as a less expensive radio that has been properly connected to a really fine antenna. Beginning to get the picture? We'll spell it out even a little more clearly: An antenna is the single feature of a radio system that can supply either clearly superior radio performance or make the radio nearly useless.

Getting the right antenna will be the single most important part of your Citizens Band radio shopping, whether you want one for a mobile unit or a base station, because no matter which kind you're after, you've got just a certain amount of transmitter power to utilize: 4 watts. The only way you're going to get all your money's worth out of your new radio will be to hook it up to an antenna that perfectly suits your needs and that will fully radiate all the power that the transmitter sends to it.

The basic prerequisites? They're not much different from the ones you'd more than likely apply to the process of buying a new car, or a new television, or anything else. You're looking for high efficiency, excellent reliability, and low cost. Keep in mind that the cost factor is relative. A CB radio set's range is only as good as its antenna, so you might as well start off with the very best one for your purpose.

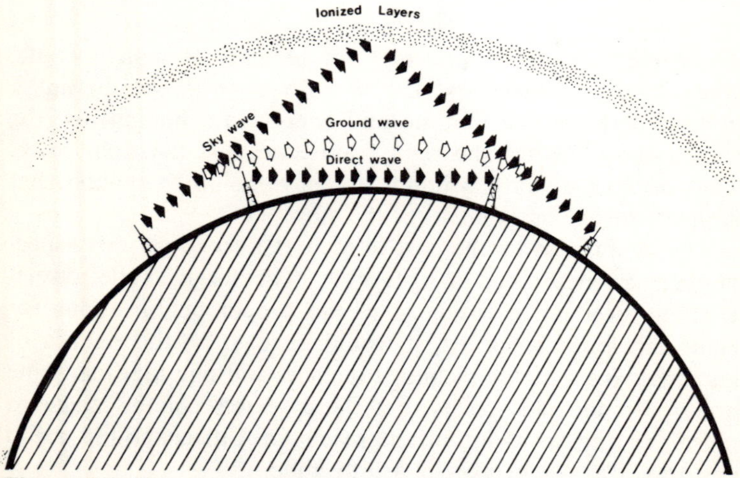

Figure 7-1. This is an artist's representation of the three kinds of radio waves. Direct waves go directly from antenna to antenna, and will tolerate few obstructions. Ground waves have a greater range because they follow the curvature of the earth and are not hindered as much by natural obstructions as are direct waves. Sky waves provide the greatest range because when the atmospheric conditions are just right, they can bounce off the bottom of the earth's ionized layers and come back to earth miles from where they were sent.

Steve Roecker

To fully understand why antennas play so important a part in Citizens Band radio performance, perhaps it would be useful to understand a little about the radio waves your CB set generates, what they are and how they work.

Basically, there are three different types of radio waves, and the direction they take as they travel through the atmosphere and across the surface of the earth depends at least in part on what their frequency range is. The most common sort of radio waves are called ground waves. They're generated by transmitters operating in the lower regions of the radio spectrum, and are called ground waves because they tend to follow the curved surface of the earth. They're the sort of waves that your car's AM radio receives. Because they follow the earth's curvature and are not as seriously affected by natural obstacles in their path as are certain other kinds of waves, they can often be heard over distances far greater than mere line of sight. If you've ever driven along in the night and listened to Wolfman Jack coming in over XERB from distant Tijuana, or to the old Grand Ole Opry show when it came from WSM in far-off Nashville, you've experienced the ground-wave phenomenon. You won't experience it often with a Citizens Band radio, however.

The second kind of radio wave is the direct wave. It is generated by transmitters working a little farther up in the frequency range of the radio spectrum. Because radio waves tend to follow the earth's curvature less and less as their frequency increases, direct waves are just that: direct. They follow line of sight, and can roughly be compared to the kind of wave a television broadcasting station generates. The direct wave is a fine carrier of information, there's no doubt about that, but to put any kind of sizable obstacle in between a direct-wave transmitter and receiver is to completely cut off electrical contact. Citizens Band radio, because of where it falls in the frequency spectrum, uses direct-wave transmissions, and that's one of the reasons CB's range is rather short, limited in most cases to line of sight.

There are times, however, when CB's range increases dramatically and appears to be independent of natural obstacles, and that's because of the third kind of radio wave, the sky wave. The sky wave is nothing more than a direct wave given off by the antenna at a much steeper angle of incidence than a direct wave. Instead of being useful and heading off in the general direction of another antenna, it takes off for the clouds, never to be heard from again—unless that atmospheric oddity known as skip happens to occur.

Skip is a condition that carries certain kinds of radio waves over long distances, and it usually lasts for a limited number of hours and can occur only when the atmospheric conditions are just right. Simply put, it means that the sky waves generated by a transmitted antenna and radio bounce, or skip, off the bottom side of the earth's ionospheric layer on their journey into space, and head back toward earth. Skip is the phenomenon that enables radio signals to travel around the world, and just how long skip is in, and how effective it is, depends on how the layers of the ionosphere are arranged. That arrangement can be from 50 to 400 miles out into space and varies with the change of season and even between day and night. You can't try to outguess skip, and neither will FCC rules and regulations allow you to use it, although many CBers do. Its variations usually can be predicted, but sometimes they can't be. It's best to keep in mind the fact that the FCC considers Citizens Band radio a short-range communications tool, and the fact that the use of skip to increase your radio's range is illegal and could result in a citation from the FCC.

You'll recognize skip when it's in because to get any quiet out of your radio you'll be tempted to turn your squelch control up so far that few other signals will be able to get through. Skip presents a tremendous amount of interference in the form of whistling, singing sounds and sometimes it also can effectively cover your radio signals so that nothing at all seems to get away from your antenna. About the only thing you can do when skip is in is to try to ignore it and not use your radio unless you absolutely have to. We should mention, however, that if you can bring yourself to listen to all the interference you'll hear, there's nothing at all wrong with listening to other CBers who, against the wishes of the FCC, are attempting to use skip. You can often hear fascinating conversations, carried on in the most interesting regional accents, from some of the most colorful, far-flung places in the country.

We've talked a bit about the behavior of radio waves, but we haven't spent much time describing just what a radio wave is. Well, you'll never see one, and you'll not feel one, either. Simply put, a radio wave is an unseen, unfelt electromagnetic presence that travels at the speed of light and is sent through space by the energy delivered to an antenna from a transmitter. Different kinds of antennas radiate radio waves in different patterns, but for starters visualize a nice, placid pond, and think about what happens to its smooth surface when you drop a pebble into it. A series of concentric rings immediately erupts from the spot where the pebble entered the water, and the rings move out and away from the center of the disturbance, getting weaker and smaller and farther apart as they travel. That's how radio waves work, at least when they're transmitted from a simple, vertical antenna. They go outward in all directions from the antenna, and they get weaker as they travel.

Because of the special characteristics of the waves emitted by Citizens Band radio transmitters, strange things can have a great overall effect on how far CB radio waves can be sent and the distances over which they can be received. For instance, the range of any given radio/antenna combination will be considerably longer over water than it will be over flat ground, and the range of that same unit will be considerably longer over flat ground than it will be in a metropolitan area. Additionally, an antenna's height, location, and type all have considerable effects on range and purity of signal.

Perhaps a working definition of the term *antenna* would be useful here. An antenna is a device, usually composed of metal wires and rods, that is hooked to a transmitting or receiving radio and either receives electromagnetic waves or radiates them into space.

Not just any old hunk of metal and wire can function as a proper antenna, though in times past such diverse things as window screens, windmills, and old combine harvesters have been used to send and receive radio signals. Antennas usually are made from thin stainless steel rods, aluminum tubing, or fiberglass—and if they're made of fiberglass they've got some metal in them, because fiberglass does not conduct electricity, and an antenna must be able to do that. An antenna can consist of a single vertical element, with the transmitter's power fed to it from its bottom end, or it can consist of a veritable forest of wires, tubes, rings, and cross members. But no matter what the antenna consists of, one thing is certain: If the antenna is a good performer when it's kicking a signal out, it will be a good performer when there is an incoming signal to snag.

For an antenna to be a good performer it must be the right

length. That is, its length must have a direct mathematical relationship to the wavelength of the radio waves it is designed to broadcast and receive. There are two different kinds of lengths to confuse the issue, electrical and physical. Citizens Band radio has been assigned by the FCC to a radio frequency known as the 11-meter band. In an earlier chapter we discussed the fact that Citizens Band radio works in a frequency range of about 27 million cycles, or megahertz, per second. That translates to mean that a Citizens Band radio wave will travel approximately 11 meters, between 36 and 37 feet, during a single cycle, hence the 11-meter-band designation. Now, no antenna, no matter what it's for, will be able to do the job it's supposed to do unless its physical or electrical length is the same as, or is a multiple of, the radio signal's wavelength. This means that as far as CB radio is concerned, an antenna with a length of 36 feet probably would be just about ideal, even if its length would be somewhat inconvenient. An antenna 18 feet long, half the length of the CB radio wave, would also work very well, and so would an antenna 9 feet long, a quarter of the length of a CB radio wave.

It should not be inferred from the above that antennas cannot be shorter than 9 feet in physical length, because they can be. But when they're shortened, their electrical length must remain at about 102 inches, 9 feet, a quarter-wave length, and to accomplish this a device known as a loading coil is used. A loading coil is a coiled length of wire that is applied to a shortened antenna's top, middle, or bottom, and is designed to make the antenna appear to be the proper electrical length to incoming or outgoing radio waves. In principle it's rather like the capo used by guitar and banjo players; it's an add-on that effectively changes the length of the element it is applied to. A capo works to artificially shorten a guitar string; a loading coil works to artificially lengthen a radio antenna. Shortened antennas work well and indeed, are far and away the most popular for use with mobile stations. We'll talk about them in greater detail later, but right now you should know that there is yet another characteristic, this one electrical, that also affects the performance of a radio/antenna combination.

The name of that characteristic is "impedance," from the verb *to impede,* which means "to interfere with or slow the progress of." Impedance is the electrical opposition a given circuit presents to a flow of current. Impedance is rated by unit measurements called ohms, with a small number of ohms representing a small amount of impedance, and a large number of ohms representing a large amount of impedance. Having an antenna, a transceiver, and a batch of connecting coaxial cable that are all of the same impedance is just as important as making sure that your stereo speakers and your stereo receiver/amplifier have the same impedance rating. In either case, if there is an impedance mismatch somewhere along the line, a portion of the flow of electrical energy being sent to an antenna, or to a stereo speaker, will be reflected back toward the point where it was generated. If the mismatch is a minor one, small amounts of energy will be reflected, but if it is a major one, a large amount of energy will fail to reach its intended goal. In either case, reduced performance can be the only result and when, as in the case of Citizens Band radio, you've got just 4 watts to work with in the first place, you simply can't afford to lose even a small portion of your transmitter's radio energy because of an impedance mismatch.

The amount of mismatch in any given radio/antenna system is usually expressed as a standing wave ratio, or SWR, and is the ratio of maximum voltage to minimum voltage flowing along the line between the antenna and the transceiver. A device known as an SWR meter—or "bridge" in CB slang—is used to measure the SWR or mismatch. For the best performance, either in a base station or mobile unit, the SWR reading must be kept as low as possible. The best ratio is one to one, usually achieved at the center frequency on your radio, channel 11. If the SWR is low here, it will swoop up on either side of channel 11. That's all right, as long as it doesn't go past about one and one-half to one at channels 1 and 23, where the highest mismatch would normally occur.

But to get involved in figuring out what your antenna's SWR is, you have to first buy an antenna. Before you can do that, you have to know what kinds of antennas are available, and before that, you have to decide what you are going to use the antenna for. The last is a simple decision. If you're buying a mobile station, and that is the kind of CB radio most people are buying these days, you're naturally going to look for a mobile antenna; if you're buying a base unit, you're looking for a base antenna. Simple distinction? Well, maybe, but it helps to know exactly what you're after when you go in to talk to the salesman at the CB store.

If it's a mobile antenna you're after, you've got two basic choices, and several subchoices. The basic decision is whether

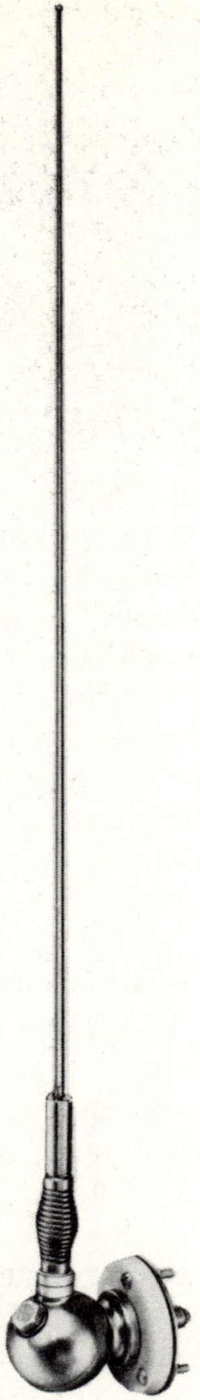

Figure 7-2. Many CBers prefer a full quarter-wave whip when it comes to mobile antennas. This is a typical one. It is 102 inches long, made of fiber-glass, and mounts on a ball socket that screws to a car's rear fender, to a van's side, or to a pickup truck's rear bumper. *Courtesy of Breaker Corporation.*

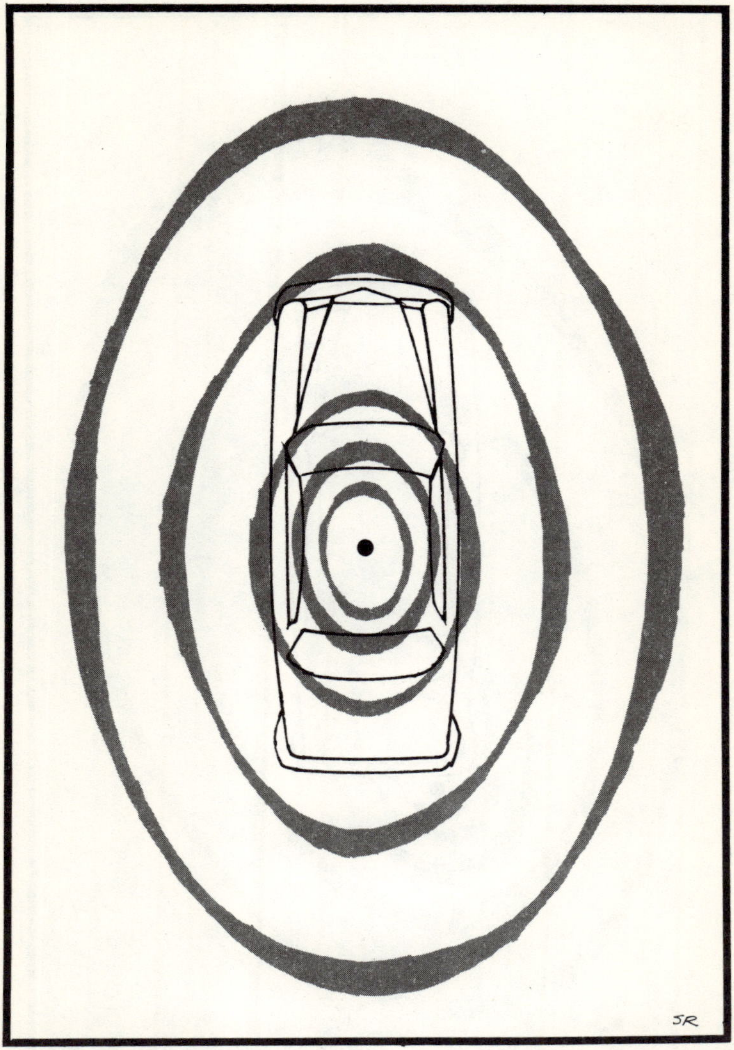

Steve Roecker

Figure 7-3. This is the sort of radiation pattern that you can expect from an antenna mounted squarely in the center of a car's roof. Overall range might not be as great as it might be with a bumper-mounted quarter-wave whip, but the distribution of the waves the antenna send out is much more equal.

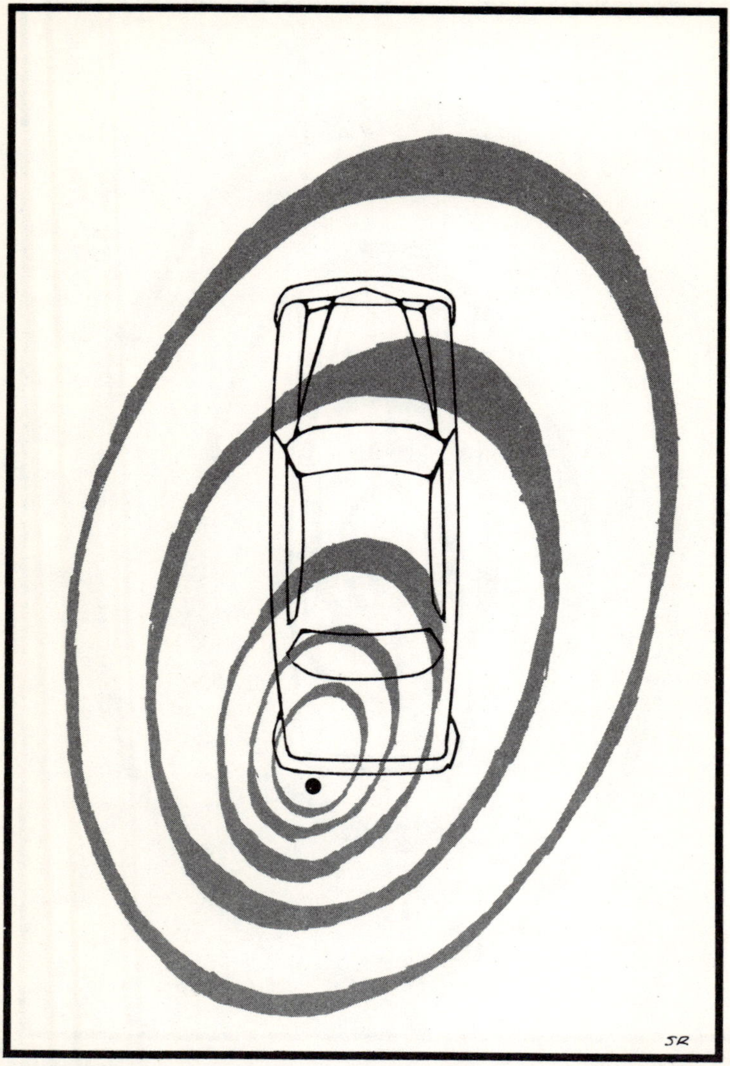

Figure 7-4. For the ultimate in range from a mobile unit, the answer is a quarter-wave whip mounted on the car's rear bumper or on a rear fender. As you can see, the radiation pattern resulting from this mounting position is a bit lopsided.

Steve Roecker

to buy a full 102-inch-long quarter-wave whip antenna so you can take a shot at knocking out your local gas station's overhead lighting fixtures every time you stop in to fill the tank, or to buy a shorter antenna, sacrifice a little performance, and gain a lot of convenience. The subchoices include deciding what kind of material you want the antenna to be made of and where you intend to mount it.

For starters, you should know that all mobile antennas are whips, a type favored for mobile use because of its radiation pattern. What's a radiation pattern? Remember the pebble we dropped into the pond a few pages back, and remember that we learned that radio waves leave a vertical radio antenna just as ripples leave the spot where a pebble disturbs a pond's surface? That's a radiation pattern. For a vertical whip, mounted properly with the proper ground, the radiation pattern usually will be more or less circular. Other kinds of antennas have vastly different radiation patterns, and we'll get to that later.

What's important now is to know that when you mount an antenna on a car, the car's roof serves as the antenna's ground, and only when the antenna is mounted in the center of the roof will the antenna's radiation of electrical signals be evenly distributed. The best overall performer would be a full quarter-wave whip screwed to the top of your car, but as mentioned, because of the antenna's length, such an arrangement just isn't practical.

Most people who use quarter-wave whips mount them at the vehicle's corner, on the rear bumper, and while this combination will produce some excellent range, it also provides a badly out-of-balance radiation pattern, with the strongest waves going in the direction of the opposite front corner and the weakest going off to the rear. This kind of mounting arrangement also provides a power loss, because the sheet metal of the car's rear fender and decklid interferes with the antenna's ability to radiate, so that only the portion of the antenna sticking up above the vehicle's roof is working efficiently.

For this reason many people opt for antennas that are physically much shorter than the quarter-wave but are electrically just as long, and, because of their short physical length, can be mounted on a vehicle in a more central position. This is a decision you're going to have to make: best overall range versus better-balanced radio-wave distribution, convenience, and unobtrusiveness. Shorter antennas just naturally are sim-

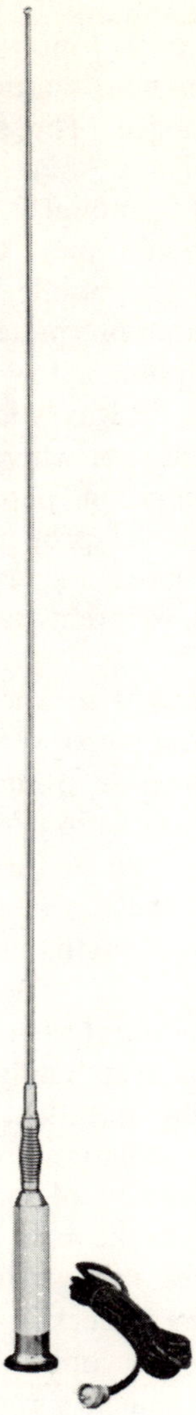

Figure 7-5. This is a roof-mount antenna. It's nearly identical to a trunk-mount antenna; the main difference is that this one doesn't have a mounting foot that can be clamped onto a piece of sheet metal. *Courtesy of Breaker Corporation.*

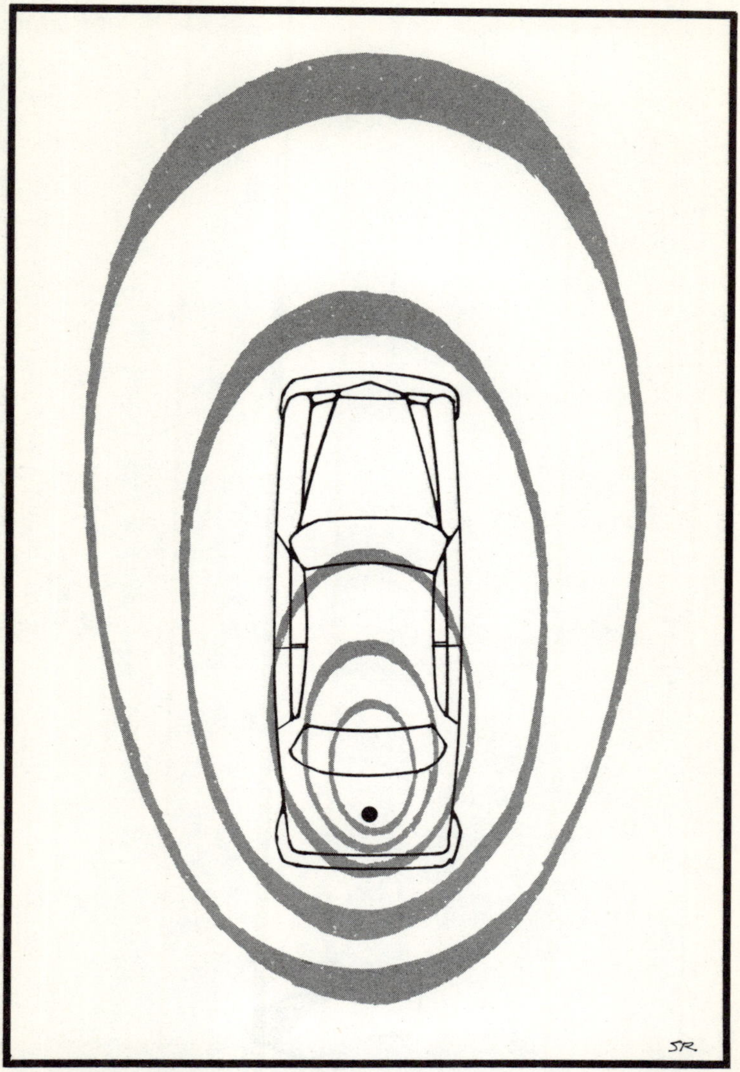

Figure 7-6. Some CBers feel that the best compromise between drilling a hole in a car's roof to mount an antenna there for best signal distribution and mounting a quarter-wave whip on a rear corner for the best range is a trunk-mounted antenna. Range remains good with this kind of mounting position, radio wave distribution remains semibalanced, and no holes are drilled in the car's sheet metal.

Steve Roecker

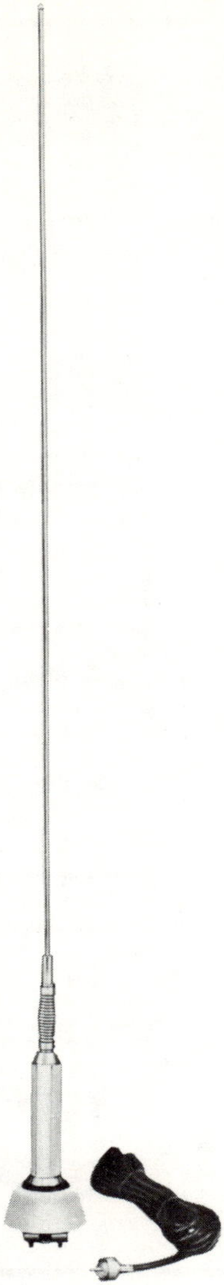

Figure 7-7. This is how a typical trunk-mount antenna looks. This example is base loaded and uses a stainless steel whip. *Courtesy of Breaker Corporation.*

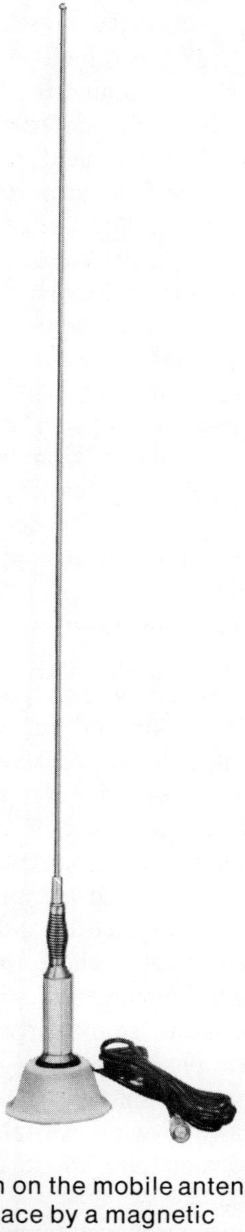

Figure 7-8. Another variation on the mobile antenna theme. This one is held in place by a magnetic base and can be put on a car's trunk lid or on its roof. *Courtesy of Breaker Corporation.*

pler, more convenient, and less obtrusive. And in most cases, they perform nearly as well as the long ones.

As mentioned, the ideal position for any mobile antenna is smack in the center of the vehicle's roof. While this certainly is practical with a short antenna, the fact that it usually means drilling a hole in the car's roof causes most people to shy away from this solution. The next best position, and one that results in very little performance loss when compared to a roof mount, is at the leading edge of the rear decklid. Nearly every antenna company sells some sort of clip-on rear-deck-mount antenna that can be used without drilling holes in the car's sheet metal, and many people favor them.

There are many other kinds of mounts for shortened antennas, and most of them present performance disadvantages. Some of them present some very real practical advantages, however, so we'll talk about them briefly. Suppose you have two cars and just one mobile unit, and you want to use your radio on both cars. The thing to do is to get either a clip-on antenna that can be attached to a rain gutter, or a magnetic-mount antenna that will stick nearly anywhere you wish to place it. In either case you run the coaxial cable that connects the antenna to the radio in through the car window or through the door's weather stripping, and you've got an antenna that can be dismounted quickly and easily. Or suppose you've got a truck. Antennas now being marketed can be attached to the big outside mirrors, and a popular configuration is to use a pair of antennas, because the use of a second properly installed and matched antenna will almost always provide very real increases in radio range. Dual antennas are also available for automobiles, and are mounted on either side of the trunk lid. Suppose you live in a city, where there is a real danger that a Citizens Band radio antenna will stick out like a sore thumb, pointing your car out to rip-off artists as a vehicle that contains an expensive piece of radio equipment. The solution here is an antenna that masquerades as a regular AM radio antenna, but actually provides signals for both the AM radio and the CB radio, and lives where the old AM antenna was originally mounted.

What should your mobile antenna be made of? Fiberglass or stainless steel. One is as good as the other, except that a steel antenna isn't quite as visible as a fiberglass antenna, and a fiberglass antenna tends to stand up at attention at high speeds, while the steel tends to bend.

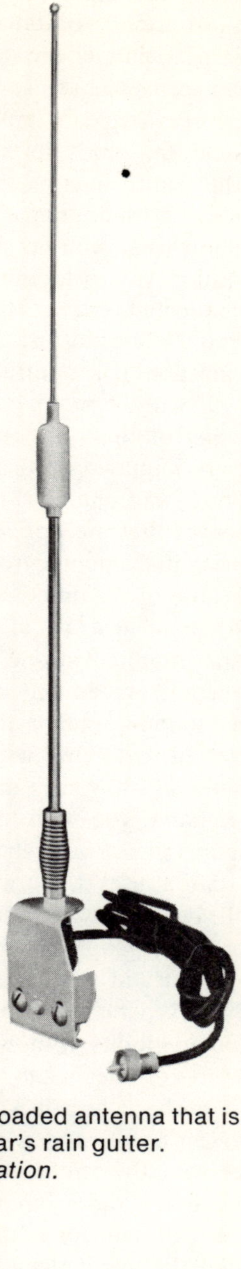

Figure 7-9. This is a center-loaded antenna that is built to be clamped onto a car's rain gutter. *Courtesy of Breaker Corporation.*

All shortened antennas intended for use aboard a car or truck use some sort of loading. Some have the load in the top; some have it in the bottom; some have it in the middle; and some, the fiberglass antennas, are continuously loaded. The location of the load seems to have some effect on the antenna's radiation pattern; for example, a top-loaded fiberglass antenna with a tunable steel tip sends its signal out from a higher point than any other short antenna. Since antenna altitude has such a direct effect on radio range, it is important to keep load location in mind when you go shopping.

If it's a base station you're looking for, things begin to get a little more complicated. Again, the first thing you should decide is how you intend to use your radio, because the answer to that question has a direct bearing on just which sort of antenna will work best for you. This is so for several reasons, and the first of them is polarization.

Antennas that are mounted vertically emit radio waves that are vertical in nature, and antennas that are mounted horizontally emit radio waves that are horizontal in nature. True, you can send to a horizontal antenna from a vertical antenna and vice versa, but because of the difference in polarization, vertical versus horizontal, there will be a definite performance loss because an antenna mounted in one plane best picks up radio waves if they're sent from an antenna mounted in the same plane. Now, every mobile antenna available is a vertical one, so if you're interested in talking with and listening to mobile units, thet best antenna for you is one that is vertical.

If, on the other hand, you wish to talk to a particular base station—say, one belonging to a friend a few miles away—and that other station uses a horizontal antenna, your best bet probably will be to buy a horizontal antenna. By and large, though, for most people a vertical antenna probably is the best choice, at least in the beginning of their CB experience, because of its even radiation pattern, ease of installation and tuning, and compatibility with mobile units.

Another factor to consider when you're looking for a base-station antenna is a characteristic known as gain. Gain translates to mean power boost, and while there isn't an antenna in the world that actually can increase the power of a radio signal in the way that a radio circuit does, certain antennas, because of their design, can focus their radio waves in such a way that the signal they emit does appear to be a good deal stronger than the signal being sent to the antenna by the transmitter. This gain is most often found in antennas that are expensive and highly directional in nature, but it can be engineered into omnidirectional antennas as well. This is done by altering the angle of radiation of sky waves so that very few radio waves are emitted at steeply climbing angles. Instead, they're focused out at a flat angle, so they have a much greater chance of reaching other antennas. They also have a smaller chance of being picked up by skip. Omnidirectional antennas that can effectively double the power of a transmitter's signal are not uncommon, nor are they illegal, since the FCC limits only the transmitter's output, not an antenna's ability to direct a signal.

Omnidirectional antennas are the most commonly seen type, and for good reason: They put out the most evenly balanced pattern and, as noted, are best for use with mobile units. The ground-plane antenna is the most typical type. It usually comes in the 9-foot quarter-wave length, although it also can be found in 18-foot, half-wave lengths. It uses a straight-up radiating element and an artificial electrical ground plane that simulates earth's ground even though the antenna may be mounted several wavelengths above the ground—hence the name. The ground plane usually consists of three or four isolated radial spokes or elements projecting from around the bottom of the upright element. For proper performance the radial rods either have to be a quarter-wave in length or, if shortened, electrically loaded and about a quarter-wave's distance above the surface of the earth. Where gain is concerned, it should be noted that the doubling of a signal's intensity doesn't necessarily mean that range will be doubled. It won't be, but range will increase perceptibly.

Another common type of base-station antenna is the coaxial. It is particularly useful when the chosen installation location provides no room for a ground-plane antenna's long radial elements. A coaxial antenna is usually about 18 feet, a half-wave, in length. The top half is a quarter-wave whip, and the bottom part, also a quarter-wave in length, is a hollow skirt that has been isolated from the topmost section. The lead-in wire feeds in through the skirt, which acts effectively as a ground plane. In all a very effective arrangement, but the possibilities for gain aren't nearly as good with a coaxial as they are with a ground plane, although a good coaxial does keep the angle of radiation of sky waves as low as possible. A drawback to the coaxial antenna is that it exhibits a high

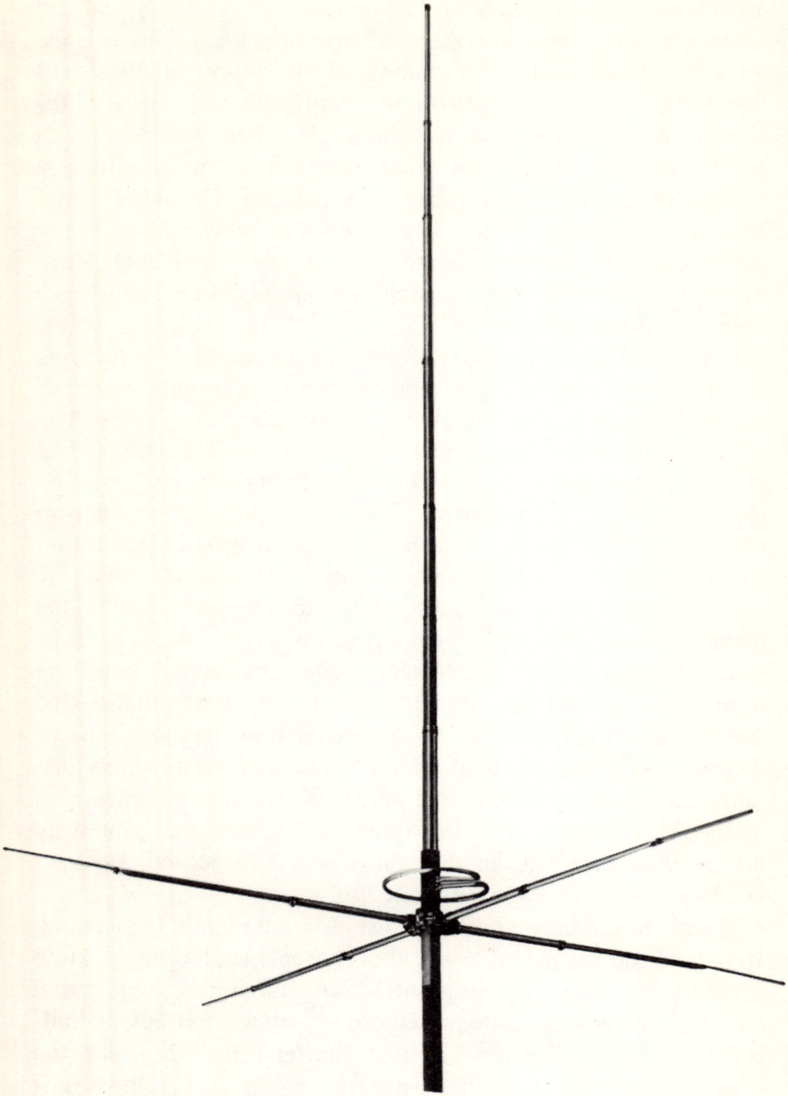

Figure 7-10. This is what a typical ground-plane antenna looks like, a single vertical mast and four radiating ground-plane elements. *Courtesy of Avanti Research & Development, Inc.*

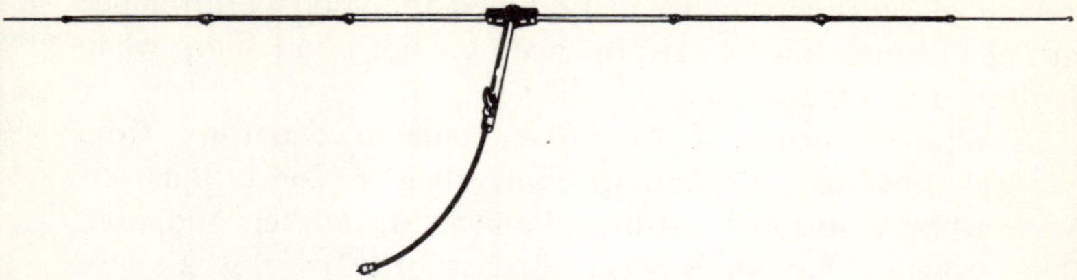

Figure 7-11. A typical dipole antenna. This one is adjustable so that it can be used for several different wavelengths. *Courtesy of Avanti Research & Development, Inc.*

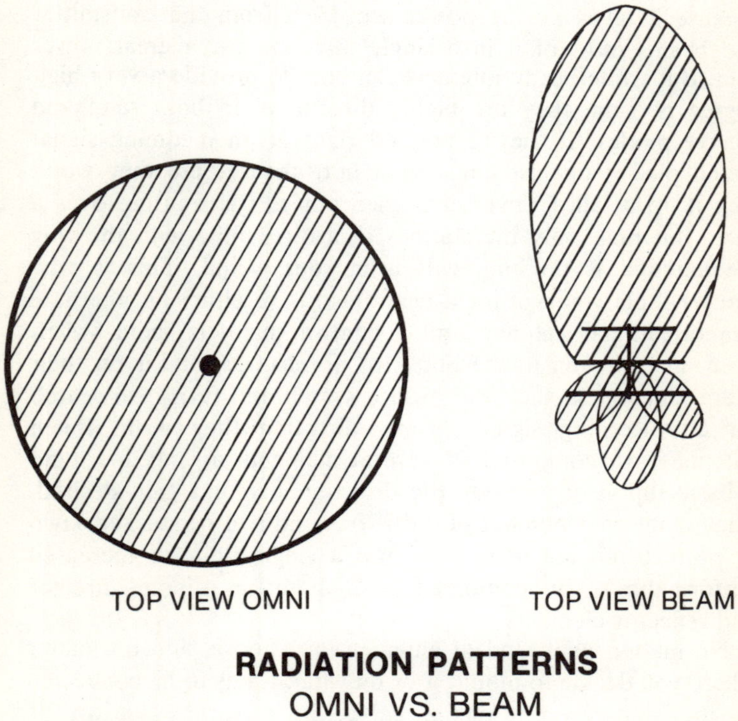

Steve Roecker

Figure 7-12. This is an illustration of just how different the radiating patterns of an omni-directional antenna and a beam antenna are. The omnidirectional antenna, represented on the left, radiates in a circle, and the beam antenna, represented on the right, both transmits and receives in just one specific direction.

degree of sensitivity and can be tuned for good performance at one channel, but its performance may not be as sharp when it's tuned to other channels.

A dipole antenna is the most common departure from vertical antennas. It's nothing more than a pair of quarter-wave whips connected end-to-end and fed from their midpoint. The radiating pattern is changed drastically by this kind of feed arrangement; from above the antenna as it lies flat in its usual horizontal position, the pattern looks more or less like a huge figure eight.

If lots of gain is what you're looking for and you're interested mainly in communicating between two fixed stations, what you want is a beam antenna. A beam antenna is so called because it takes all the power sent to it from the transmitter and beams most of it in a single direction for a great power gain. But while beam antennas can and do provide a very high degree of gain, they are highly directional in both send and receive modes. If they're pointed right at an incoming signal they'll pick it up just fine, but if not, chances are they won't pick it up at all. Conversely, when you're transmitting over a beam antenna, only the stations in a direct line with the way the antenna is pointing will hear your signal. This is why many people who opt for a beam also have either an alternate omnidirectional antenna and a control that lets them switch from one antenna to the other, or an electric rotator on their beam antenna so they can push a switch and turn it to search for an incoming signal in any direction they wish. Beam antennas are composed of several elements arranged around a basic dipole to provide the directionality and gain desired. They come in a number of variations and can be stacked (two or more beam antennas built into a single unit for increased performance), or combined with parasitic dipole director and reflector elements.

No matter what kind of antenna you choose, and no matter where you decide to mount it or on what, it has to be connected to the transceiver. Coaxial cable, in a variety known as RG-58/U or RG-8/U, is used. This usually is a black, plastic-coated wire about ¼ inch in diameter. It is composed of a stiff copper or steel inner wire that has been wrapped with insulation. On top of the insulation is a braided copper shield, which is topped with the black plastic coating to protect it. The important thing to remember about coaxial cable is that the more of it you use, the more power is lost, so you want

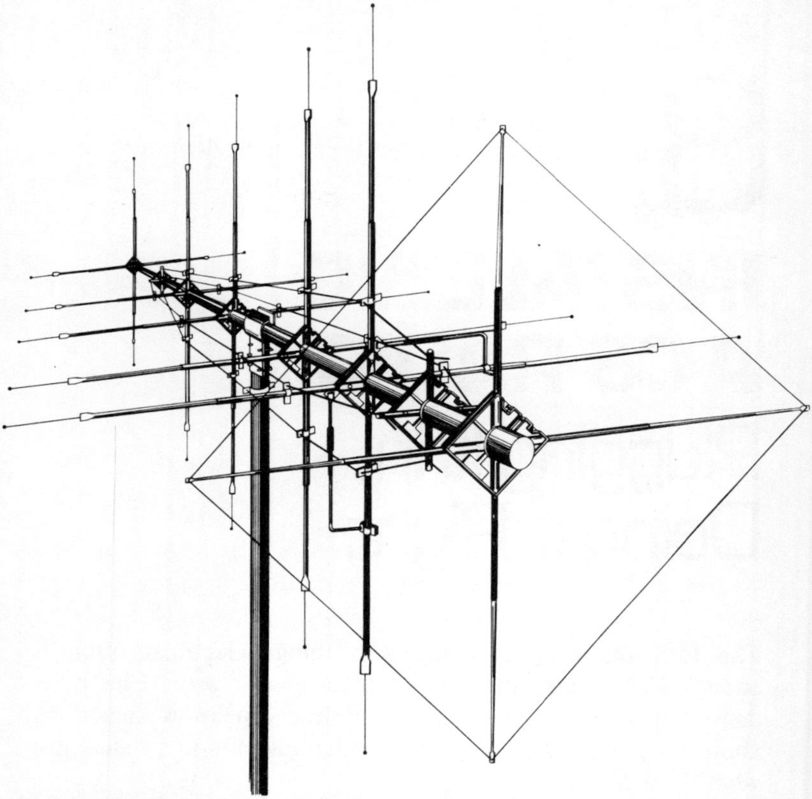

Figure 7-13. Beam antennas can get bigger than you might think. This one, an Avanti Moonraker 6, weighs nearly 40 pounds and is 31.5 feet long. *Courtesy of Avanti Research & Development, Inc.*

your cable as short as possible. It's something to keep in mind when you decide where to install your station and where to mount your antenna, and is especially critical with base installations.

There's just one last thing to keep in mind when you set out to buy your CB radio's antenna. It is going to live on the roof of your house or on your car, and in either case you want to be able to consider the installation a permanent one. So before you make your purchase, be careful in your selection, buy a well-known brand, and do the installation right. If you slip up in any of these three areas, you face the possibility of having to do the installation all over again, maybe with another antenna, and if you're like most of us, you don't need to waste the time or money.

8

INSTALLING A CB RADIO:

Plugging It In, Doing It Right

The idea of messing about with things electrical often is enough to put most folks off—and for good reason. Electricity doesn't flow, as water does, and unless you know something about its habits, it's sometimes rather difficult to predict just what it might do.

When you're fooling around with things that run on electricity, it's easy to make a mistake. A misstep can shock you or burn you, blow out components of the electrical circuits in your house or car, and maybe even bring total ruin to an expensive piece of equipment. So you'll be delighted to hear that there is nothing complicated about the installation of Citizens Band radio equipment—indeed, if you follow the installation instructions that come with your equipment, you'll find it pretty hard to get yourself into trouble. You do have to follow the instructions, though, and follow them explicitly.

The first thing to keep in mind is that the way you carry out the installation of your equipment can make you or break you. If you do a sloppy, hurry-up job, the performance of your station can suffer considerably, and if you fail to sit down and think about the hows and whys of the installation before you proceed on it, you stand a strong chance of having to do the job all over again with all new equipment, or, at the very least, wasting a lot of time while you figure out just what it was that you did wrong the first time around.

If half the job of a radio installation is getting it done in the most elegant manner, as cleanly as possible, the other half of the job is choosing correct mounting locations for the two major components of your radio system—the transceiver and the antenna. That's not hard, either; it just takes a little common sense.

Let's start with the installation of a mobile unit, since sales figures tell us that mobile units are the sort of CB radio that most people are buying these days. The first step, and one of major importance, is to make certain that you have everything you need before you get started, and that everything is within easy reach. Just what *will* you need? You'll need an electric drill, a variety of high-speed drill bits (just which size bit you will want will depend on the size of the various mounting screws for the radio bracket and antenna), a small center punch, a hammer, an extension cord 10 to 12 feet long, a roll of racer's tape (also known as duct tape), a Phillips-head screwdriver and a regular screwdriver, both medium-size, an Allen wrench, a pair of diagonal cutters, and a pair of regular, garden-variety pliers. It should be noted that this is a basic tool list. Every installation presents different problems, because every car is different. You'll want to take a little time to thoroughly read your radio's installation instructions and do a complete reconnaissance of your car to size up the job ahead of you and see just what is called for.

You can do that when you decide on a mounting location. For this you'll have to hoist yourself up out of your easy chair and get out to the garage, where your car lives. It might be helpful to take your radio and its mounting bracket with you when you go. The object of this little exercise is to determine the most felicitous mounting position for the radio. That, of course, depends a great deal upon how the interior of your car is arranged; we will not attempt to recommend a "best" mounting position, since every car is different (as is every driver), but we will tell you what to look for so you can make the best choice yourself.

For starters, the radio must be located so that it presents no interference at all with the operation of the vehicle. You've got to choose a spot that will not allow the radio's housing to interfere with the operation of any of the control pedals or the parking brake, or with the gearshift lever, if you have a car with a standard four-speed transmission.

Second, the radio has got to be in a place that is completely

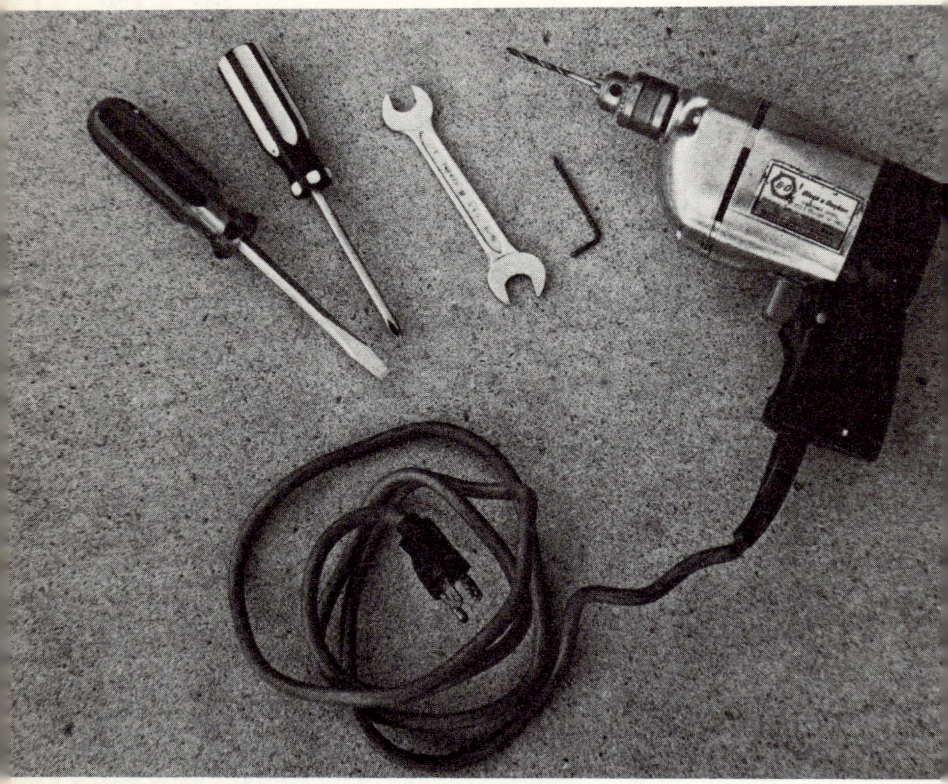

Figure 8-1. If you're going to do a mobile installation, you can count on needing a few basic tools, including an electric drill, a Phillips-head and a regular screwdriver, an Allen wrench, and perhaps an open-end wrench or two. Every installation is different, though, so thoroughly survey your own job before you begin and line up everything you need. *Photos by Steve Roecker.*

out of the way so that it won't be bumped and joggled by you or your passengers, but not so far out of the way that you as driver of the vehicle cannot reach it with your lap and shoulder belt in place. (Editorial: Good CBers *always* wear the lap and shoulder belts whenever they're in a car.)

Third, you've got to find a mounting position that offers firm support to the mounting bracket and the load it will carry. No road is as smooth as glass, and no car can iron out every bump. If you attach your radio's mounting bracket to something flimsy, it probably won't be long before the weight of the radio, negligible though that may be, weakens the mounting surface and allows the radio to bounce and flap with every alternation in the road's surface. At best this leads to the annoyance of having to constantly be tightening and retightening, and at worst it could lead to your radio's getting so loose that it bounces itself right out of its mounting holes, possibly doing damage to itself when it hits the car floor, and probably tearing unsightly holes in the bottom of your car's dashboard.

So take your radio and bracket out to your car and figure out where the best place to stick them will be. If you're doing the installation in a car, there are three general possibilities, and if you're installing your radio in a truck, there's a fourth possibility. If you're putting your radio into a car, you might mount it to the underside of the dash (and, indeed, this seems to be the location most folks choose); you might mount the thing on the side or top of the transmission hump (this makes sense, because it virtually assures that the radio will be within easy reach of the driver); or you might mount the radio underneath the driver's seat (this position is favored if the two previously mentioned locations are not workable, or if you think that you're in real danger of having your radio set ripped off if it's out in the open and visible. You'll find that this is the least handy location you can choose, however). That fourth location that seems to work out well for truckers involves mounting the radio either on the roof bulkhead right up at the roof's leading edge where it intersects with the windshield glass, or on the motor cover that bisects the cabs of the large, cab-over diesels.

Once you've decided on a mounting position for your radio, you've got to decide on a mounting position for your antenna, and there too you've got several ways to go. You can choose a roof mount, a trunk-lid mount, a cowl mount, or a rear-bumper mount. If you've read the chapter on antennas, you already may have made up your mind about which sort of mount is best for you; if so, fine, and if not, now's the time to choose your spot and buy the appropriate antenna.

Once you've made your choice and have the antenna in hand, it's time to proceed and to do a proper, professional-appearing installation job. You'll want to start by first installing the antenna, leaving the radio completely alone for a while. Why? For starters, you'll want to be troubled with just one set of parts at a time, and with just one installation problem, but there are a couple of other good reasons for proceeding this way. First, you'll want to be able to plug the antenna into the radio as soon as you get the radio mounted on its bracket, before you hook up the power leads (we'll explain why later). Second, the length (or lack of it) of the cable supplied with your antenna could conceivably have some effect on where you decide to stick the radio itself.

So. The first thing to be done is to remove the antenna and all its parts from its protective packaging. Lay out all your tools, and lay out all the antenna pieces as well. Grab the directions that came with the antenna and use them to identify each of the pieces that came with the antenna kit; make sure you've got them all. If you don't (packaging mistakes are rare, but they do happen), don't go a step further. Put the whole mess into a box or a paper bag and take it, along with your instruction sheet and receipt, to the store that sold you the antenna. But make certain that you haven't overlooked something, that there really is a piece missing. If there is, ask that it be put right. If everything is there the first time around, you can go about mounting the antenna.

For starters, read the instructions over once or twice, very carefully. Then, if the mounting position that you have chosen requires that you drill or cut holes in your car's precious sheet metal, get the mounting hardware out and place it on the car in the position in which you want it. Using a pencil, outline where the hardware will go and mark the holes you need to make. Then take your center punch and hammer and punch yourself a little indentation at each spot where you need to drill a hole. Don't try to punch all the way through the sheet metal; these little punch marks are to help you start drilling by keeping the drill bit from jumping off the mark and running all over the car's sheet metal when you push the drill's trigger. Just go ahead and drill your holes and, following

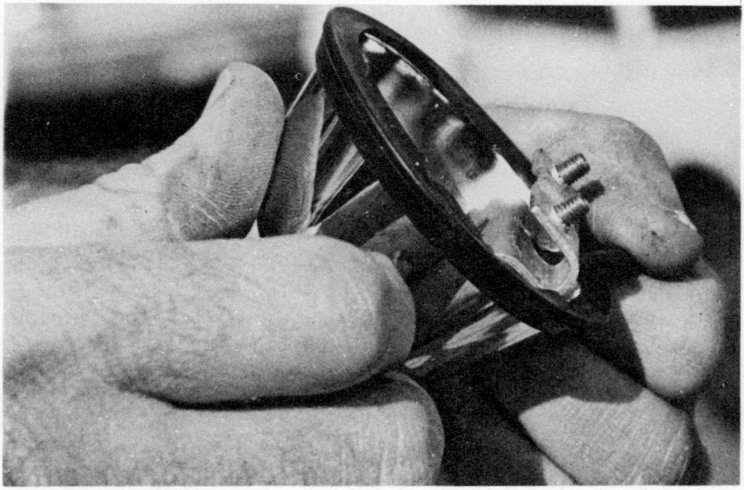

Figure 8-2. This trunk-mount was installed on a Volvo sedan, using a Johnson Messenger 130A radio and an Antenna Specialists' trunk-mount antenna. Here you see the lip on the base of the antenna. It fits over the decklid's leading edge; the two Allen screws you see are tightened down to keep the antenna in place.

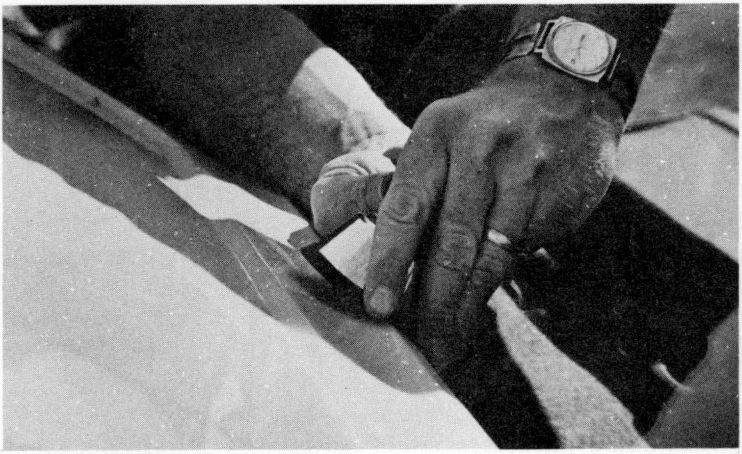

Figure 8-3. To get started, slide the base over the decklid's leading edge, thus, getting it as close to center as possible. If the lip on the antenna base isn't wide enough to fit over the edge of the trunk lid, spring it a bit with a pair of heavy pliers and a large screwdriver.

the instructions furnished you, proceed with mounting the antenna base.

If you have chosen a trunk mount, you already know that the antenna is going to live smack in the center of the trunk's leading edge, right behind the rear-window glass. It's held there by a small clip that slips over the leading edge of the decklid. The underside of the clip will, in most cases, have a pair of Allen screws that are tightened up against the underside of the decklid to keep the mount in place. For starters, back those screws all the way out and put them where you won't lose them. Then make sure that the clip will indeed slide over the edge of your car's decklid. Most of them will most of the time, but in some cases, particularly with some foreign cars, a little modification may be necessary. If this looks like it will be a problem for you, take a good strong pair of pliers and a good heavy screwdriver and spread the jaws of the clip until it will fit the way you think it should. Then slip the clip in place and, using an Allen wrench of the appropriate size, reinsert the Allen screws into their holes and tighten them down. Next step: Close the decklid and fasten on the top portions of your antenna. Just how many parts you have depends on what sort of antenna you bought. Just follow the instructions that came with the unit and you'll have all the specifics (note Fig. 8-6).

What you've got now is the mount firmly affixed, with the coaxial transmission cable just hanging there loose in the trunk cavity. If you can get the cable to go under the weatherstripping that keeps moisture out of the car's trunk, fine; if you have to run the cable over the top of the stripping, make certain that it does not firmly contact any of the car's metal body parts. This could pinch the cable and cause deteriorated performance. If you have to, take a hacksaw and cut a small channel for the cable to run through, and cover any rough edges with racer's tape. Don't try to cut a place for the cable in the weatherstripping, because if you do it wrong you'll be replacing the stripping, and you don't need that headache.

The object is to get the cable to run from the trunk compartment into the passenger compartment, and to do this you have to get it past the back seat. Take a look around, both from the trunk side and from the seat side, to see if you can find a nice, wide, easy spot to feed the cable through. If you can find no such spot, look around from the trunk side to see if there is a convenient plug that you can knock out to feed

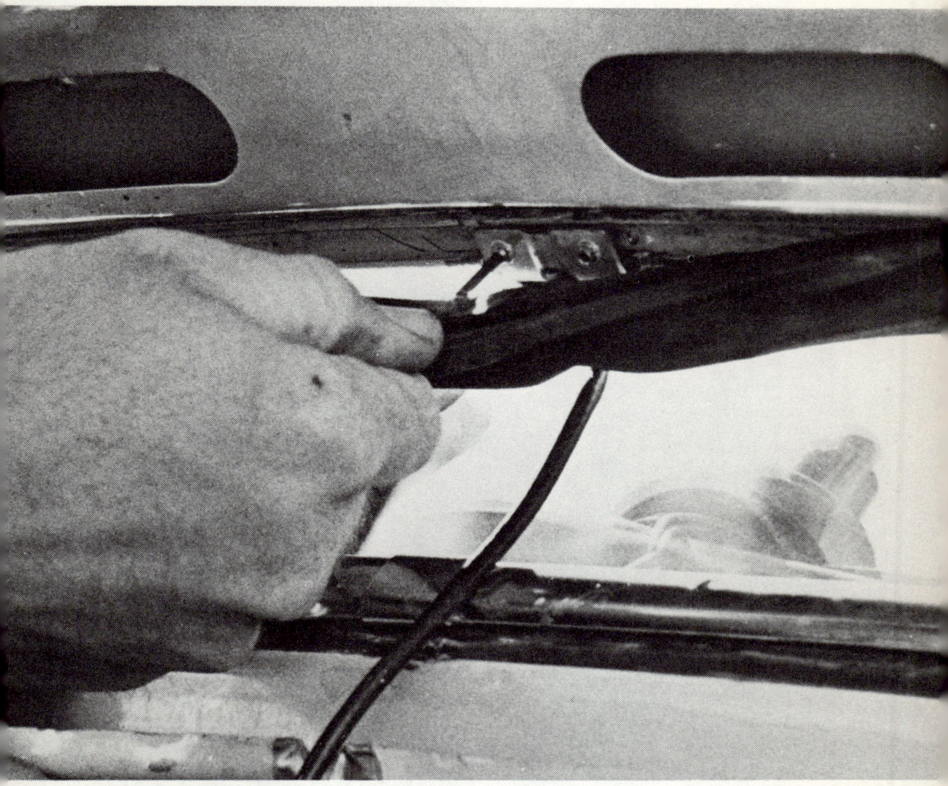

Figure 8-4. Once you've got the base in place, tighten the Allen screws down tight. As you can see in this picture, the weatherstripping on this car's decklid was accidentally pulled away during the installation. Don't let that happen if you can help it, because getting the rubber strip on most cars back where it belongs is a real hassle.

Figure 8-5. Once you've got the antenna base firmly in place, attach the rest of the antenna to it. This rustic individual is screwing on the loading coil and spring that are part of this particular Antenna Specialists unit. That done, he'll attach the whip.

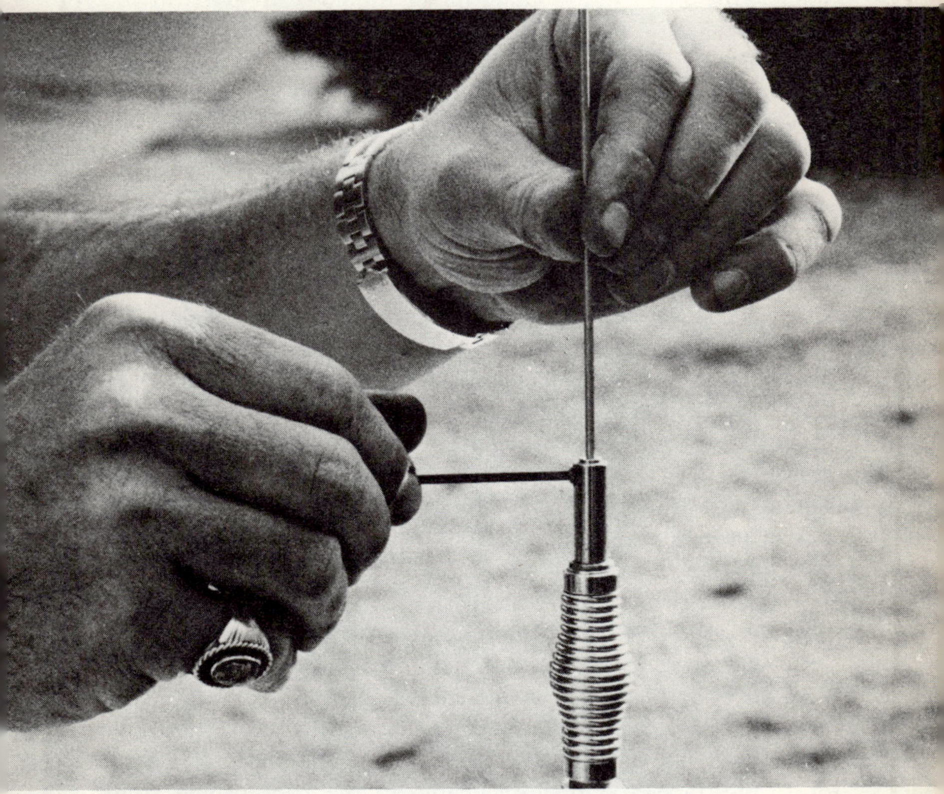

Figure 8-6. When you put the whip in place, don't tighten its locking Allen screw too much. You'll have to diddle with it some more when you go to adjust the system's SWR.

the cable through. If there's no plug and no nice, wide-open spot, you may just have to drill a hole through the bulkhead. Once you get the cable into the passenger compartment you're going to want to run it along under the edge of the carpeting or floor mats, so aim for a spot off to one side of the rear-seat back. If the radio will be living on the driver's side of the car, aim to drill your hole on the left; if it will be mounted on the right side of the car, plan on running your cable along the right. Drill a hole no larger than you need to fit the connector fitting through the bulkhead, and when you're done dress the edges of the hole with a file to get rid of any rough edges and fit a rubber grommet to the hole to keep your antenna lead from chafing through.

Now take the leading edge of the cable and find the sliding, threaded collar, the one you'll use to screw the cable to the back of the radio. Take a thin piece of racer's tape and place it around the plug assembly with the collar in its normal position. As you push the cable past the back-seat bulkhead you're going to be trying to get it through some very tight places, and without something to hold that sliding collar in position it can stay stuck in a tight spot while the rest of the cable goes its way.

Once you've decided how to get the cable from the trunk into the passenger compartment and you've done whatever is necessary to accomplish that, pull the cable through all the way into the passenger compartment and let it lie. Go back around to the trunk and make certain that the cable has no kinks or twists or tight bends in it; then take several lengths of racer's tape and tape the cable to the roof of the trunk compartment so it will stay completely out of the way and give the appearance of a good, clean installation.

Now go back around to where you've left the cable in the back seat and make sure it's got an easy, unobstructed path to the floor of the car. If you had to remove your car's back seat to get this far, you now can replace it. Lift up the sides of the floor mats or carpeting (in some cases you may have to remove a few screws and lift away a doorsill plate or two) and tuck the antenna cable in underneath the floor covering. Remember, you want a nice, straight run to your radio's ultimate location, you want no kinks or twists, and you must keep the cable away from mechanical and electrical parts, as these have a rather high potential for producing noise in the finished installation. Route the cable under the front seats, or

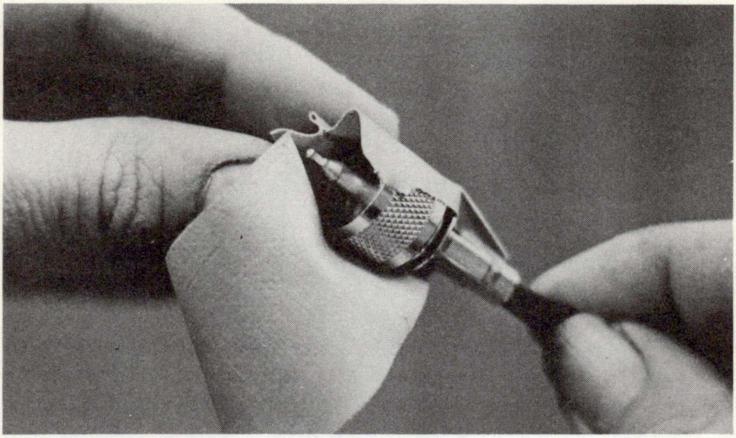

Figure 8-7. What you have now is the antenna firmly affixed, and you're ready to run the transmission cable from the trunk cavity into the car's interior. Before you start, tape the sliding collar on the coaxial plug in place so you don't lose it in a tight spot.

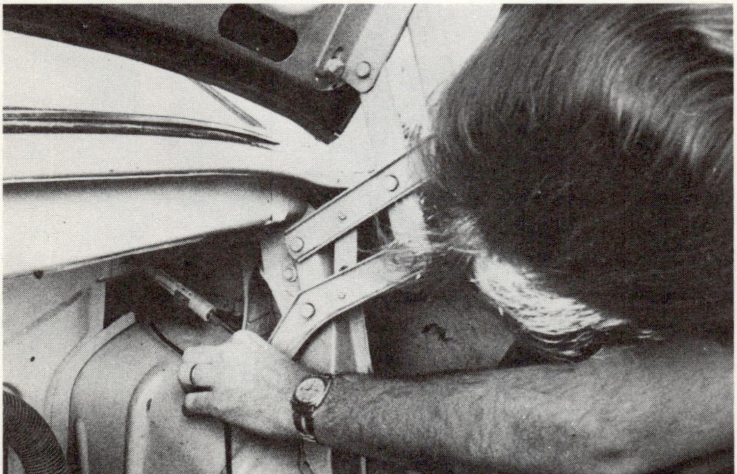

Figure 8-8. If you can find an open spot by the rear-seat back through which to push the transmission cable, good. If not, you'll have to either knock a plug out or drill a hole through the bulkhead. Make certain you feed the cable through on the side the radio will live on.

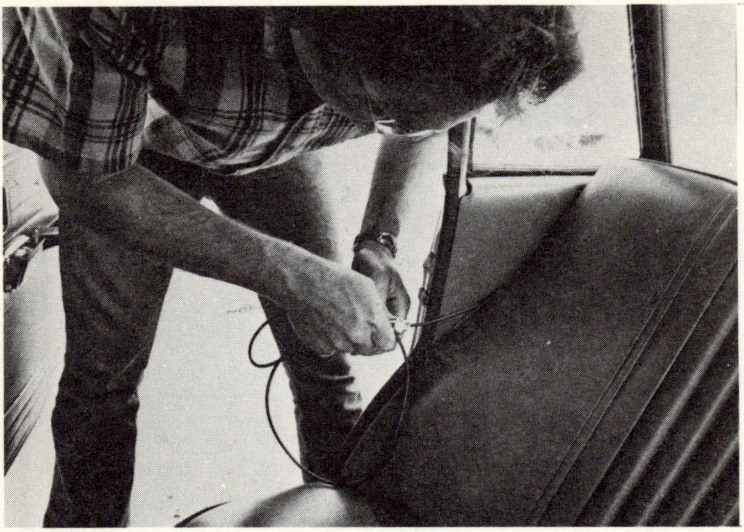

Figure 8-9. Once you've gotten the coaxial plug past the back seat, the hardest part of the antenna installation is done. Pull all the cable out; make sure that it has no twists or kinks.

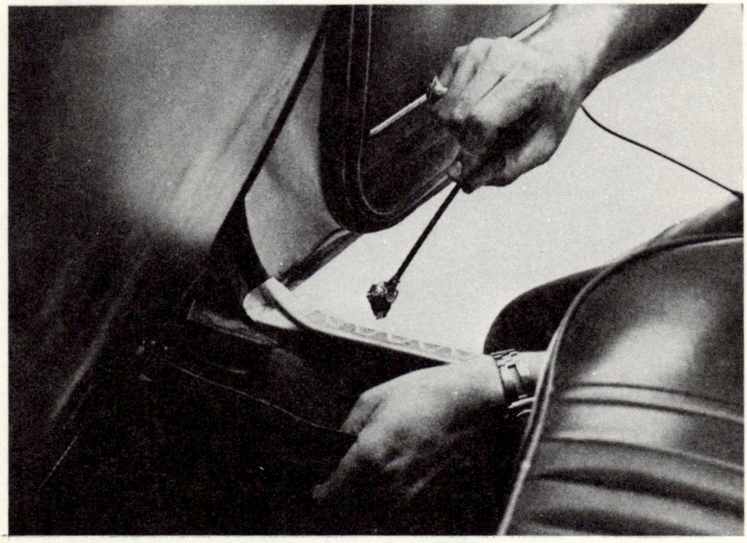

Figure 8-10. Now feed the cable to the radio's location, laying it under floor mats and carpets. You may have to unscrew threshold plates to loosen the edges of the mats/carpets.

alongside them. If you are mounting your radio under the car's dash, you'll want to run the cable underneath the floor covering in the front seat, just as you did in the back seat; if you're mounting the radio on the transmission hump, you'll want to leave the cable uncovered as it runs out from under the front seat. In either case, once you've gotten this far you can sit for a bit and have a cup of coffee, because half the job is done.

If you've chosen a rear-bumper or fender mount, the procedure for running the cable will be the same: Get it inside the trunk and run it into the passenger's compartment through the back seat. If you've chosen a roof mount, you'll want to remove the car's ceiling light fixture and, following the antenna-mounting instructions all the way, drill your hole about there. Feed your cable in between the roof and the headliner and run it down the inside of the car's doorpost. You can remove the inside covering of the doorpost of most cars to make this easy. If you've got a hardtop with no doorposts, maybe you'd be better off with a trunk mount.

If you've decided on a cowl-mount, the job is an easy one, because your CB antenna will fit in the same hole where your car's AM radio antenna now is. First unplug the antenna from the back of the AM radio and, using the appropriate wrenches, remove the antenna and set the whole thing, mast and cable, off to the side where you won't be tripping over it. Now take a good look at the replacement mast. You'll notice that it has a pair of transmission cables, one for the CB and one for the AM radio. Feed them both into the existing hole and, again using your wrenches, properly affix the antenna to the car's body. Then crawl under the dash and lead the cables into the car's interior. Plug the AM cable back into the car's radio and feed the CB lead down to the approximate location of where the radio will be mounted.

Your antenna is now in place, and in most cases you'll have more than enough cable. If by some quirk of fate you don't have quite enough, whip on down to the CB store and buy the length you'll need and two appropriate connecting plugs. And if the cable leaves you with a lot of excess, you'll want to buy a plug so that you can shorten the cable to the proper length. If you should have to put on a plug, first figure out how much cable you think you'll need, and then add just a bit and cut it with your diagonal cutters. Coaxial plugs are easy to mount, and the first step is to slide the plug's adaptor onto the cable, tapered portion last. Take a sharp knife and, about two

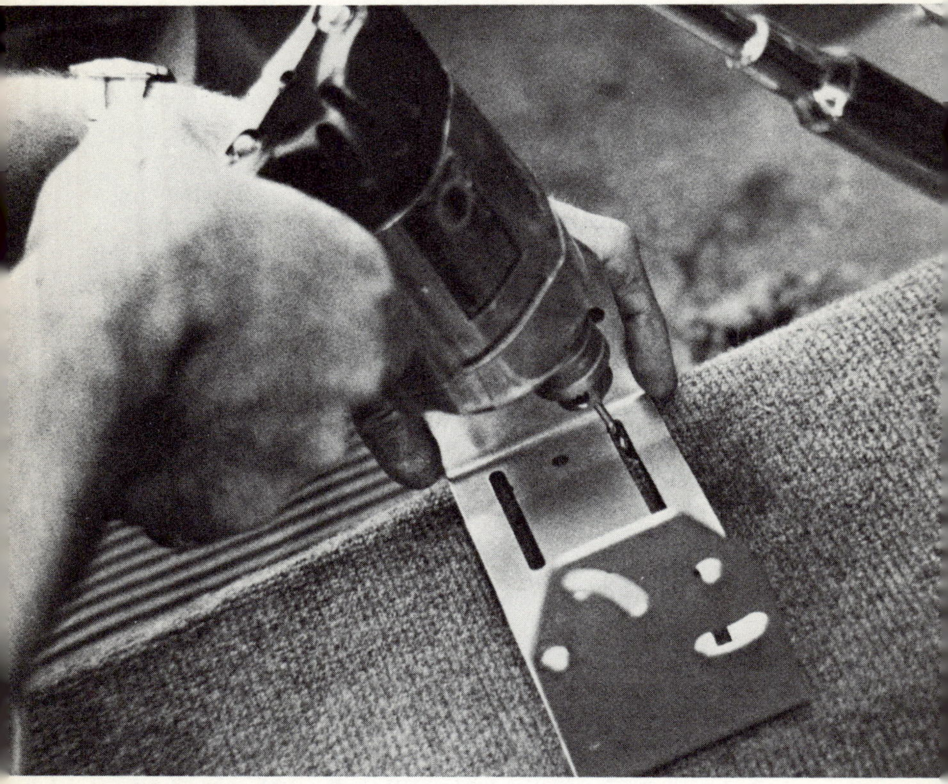

Figure 8-11. Once you've gotten the cable to the car's front-seat area and are sure the transmission line is neither far too long nor far too short, you're ready to mount the radio bracket. In this instance the car's owner has decided that a transmission hump mount is what he wants, and since he can't mark the carpet with a pencil, he's using the mounting bracket as a drill guide. If he were drilling into sheet metal he would have marked his spots with a pencil and then used a small center punch to help him get his holes started.

inches from the cable's end, cut around the entire girth of the cable all the way down to the innermost wire—but try to avoid cutting or scoring that wire. Pull the outer portions off, revealing just that inner wire. Now, about an inch back, again circle the cable with your knife, this time just deeply enough to cut the black rubber outer shield. Pull that portion of the shield off to reveal the braided copper wire that is the outer layer of the coax. Now slide the adaptor up to where the coating stops and work the braided wires back down over the adaptor's shoulder. Get all that braided wire manicured and lying flat over the shoulder and, using the diagonal cutters, clip off any excess wire that extends down over the threads. Now slip the main plug body onto the adaptor, running the bared center wire through the plug tip. Some of these are made so that they can be screwed together. Screw the pieces together so that everything fits the way it should. If you don't have the screw-together kind, you have a little soldering job ahead of you. Heat the center wire and the plug's tip with a soldering gun and drop just a smidgen of solder into the tip. Then look for the holes in the side of the plug body. Apply heat to one of these holes, and only one, until you can dribble just a taste of solder in and around that folded-back braided copper wire. Remember, use a minimum of solder and do just one hole. When you're done, you've got yourself a new, shortened antenna cable and you're ready to mount the radio itself.

The first step is to hold the radio mount to the position you want it to occupy, outline it with a pencil or felt-tip pen, and mark the holes you need to drill. If you're drilling through fabric (carpeting for instance), where a pencil mark obviously wouldn't show up, hold the bracket where you want it and use the mounting holes for a drill guide. If you're drilling through a metal dashboard, use a small center punch to mark the spots where you need holes, and go ahead and do your drilling. You of course haven't forgotten that all sorts of interesting things—wires, instruments, heaters, and such—live up inside the dashboard, so be careful not to poke the drill through so far that you damage something in there. If there's any roughness in the holes dress them with a small file.

Using sheet metal screws, fasten the radio mounting bracket to the dashboard as firmly as you can. You're now ready to fit the radio to the bracket. That done, the very next thing you want to do is attach the antenna cable to the back of the radio. You want to do this even before you attach the radio's

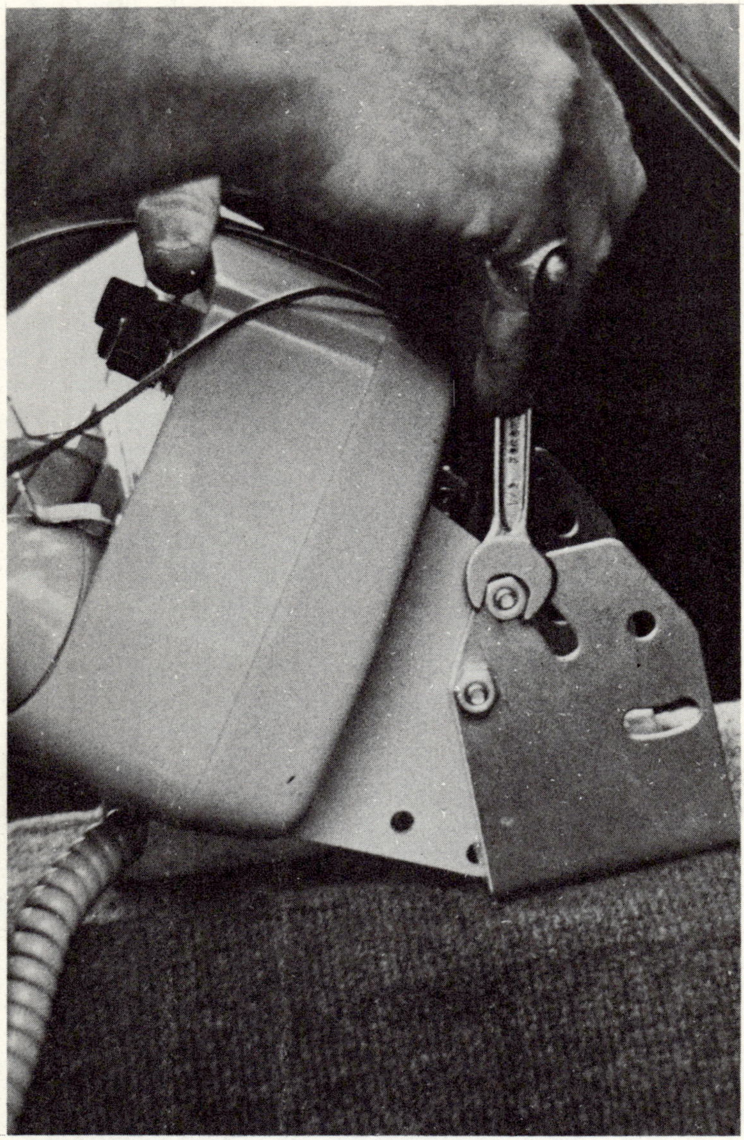

Figure 8-12. With all the bracket mounting holes drilled and the bracket firmly screwed into place, you're ready to mount the radio itself. That's what's being done here. This radio mounts to its bracket with four nuts and bolts. Many other radios are affixed to their mounting brackets with screws.

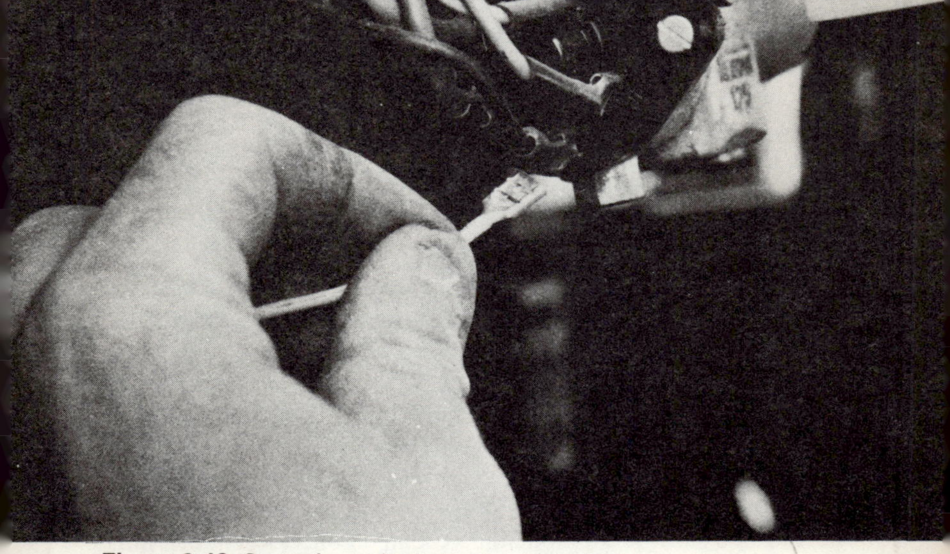

Figure 8-13. Once the radio is in place, the first thing to do is attach the antenna cable. Then hook the ground wire (the black one) to a part of the car's chassis and plug the power lead (the red wire) into the positive battery terminal, or into the hot side of the solenoid switch, or into the accessory pole of the ignition switch. The last is the hookup being used here.

power lead, because if for some reason you should hook the radio up before the antenna is tied to the set and the set's microphone is keyed accidentally, you'll stand a good chance of cooking the radio. So hook up the antenna cable the first thing.

That done, fasten up the ground wire. You'll want to run that one as short a distance as possible, and attach it to a good sturdy portion of the car's metalwork. The ground wire will be the black one. You'll also find a red wire, and that's the last one you'll have to mess with. That's the radio's power lead, and you have four choices of where to attach it: to the positive terminal of the car's battery, to the hot side of the starter solenoid, to one of several positions on the back of the ignition switch, or to the fuse block. If you decide to run the power wire into the engine compartment, where you can fasten it to the battery or solenoid, that's fine; it's a bit messy, though, and the radio will be on unless you remember to turn it off. If you accidentally leave it on when you leave the car, it will take just a few hours for the radio to drain the battery flat.

Attaching the power wire to a battery terminal can best be accomplished if the terminal has some sort of provision for this sort of thing—some do, some don't. If not, you can accomplish virtually the same thing by hooking the wire to the hot side of the starter solenoid, since that's where the positive battery cable ends. If you don't know where the solenoid is, just follow the round, plastic-covered cable from your battery to the place where it hooks onto the starter. That place will be the hot side of the solenoid. Clean it off, loosen the nut, hook your wire up, tighten the nut—and you're done.

When you are attaching bare-ended wires to a terminal the job will look cleaner if you take a look at which way the copper strands of the wire are rolled. Roll them the same direction in your fingertips and tighten them as much as you can. Then make the end into a hook, drop it around a terminal, and tighten it.

It should be mentioned that before you go ahead and hook into the positive side of your battery, make certain that your car is of the negative-ground variety. Nearly all American cars are, and so are most foreign models—but some aren't, and if you reverse the polarity, you could cook your radio. Check your instructions and see if they call for a positive or negative ground, and what you should do if the car is positive ground. Most radios these days have provision for this eventuality; just be sure you follow the directions and do it right the first time.

The way most folks seem to prefer hooking their power in is to fasten the power lead to the accessory pole of the ignition switch. This means that your CB radio will have the same power supply route that your AM car radio does. When you shut the car off and remove the key, everything goes off; there's no chance of forgetfully leaving the CB radio on and thus draining the car's battery. To find the appropriate pole, just snoop around under the dash until you discover your AM radio's main power line and where it's hooked onto the ignition switch. That point is where you want to hook up the CB lead. The ways of doing this are varied and change from car to car and from radio to radio. Some manufacturers offer a device known as a tap connector, a plastic fold-over affair that lets you easily connect your radio's power lead to any hot wire under the dash. Some cars have more than one accessory lead, and in some cars the easy way is to bypass the ignition switch altogether and hook the CB power lead to the accessory pole on the car's fuse block. It will be so marked, either on the block's surface or in your car-owner's manual. This is all pretty straightforward; just follow the directions that came in the box with your radio and you'll be all right.

So now your radio is hooked up and ready to go? Not quite. You've got one more thing to do, and that's to adjust the standing wave ratio (SWR). For this you'll need an SWR meter. First make sure that everything is off—the radio on/off switch and the car's ignition switch. Unplug the antenna lead from the radio and screw it to the appropriately marked socket on the SWR meter. Then, using the short length of coaxial cable that comes with most meters these days, hook the meter to the radio. What you've just done is hooked the antenna back to the radio, but with the SWR meter interposed in the antenna line. What you're going to adjust is the length of your antenna, and the object is to get the meter's needle to stay below 1.5:1 in three settings; channel 11, channel 1, and channel 23—the midpoint, the low, and the high of CB radio's frequency range.

Start by tuning the radio to channel 11. Turn all the power back on and key the mike. The SWR meter will have a switch and a knob, and the switch will be marked "fwd" and "ref." Go to fwd and line the needle up with the mark on the far righthand

Preceding pages
Figure 8-14. You're all but done. Check the set's SWR, making sure it is no more than 1.5:1 on channels 1, 11, and 23, and you'll be on the air, as this new CBer is.

Figure 8-15. If you're about to install a base station, know that the biggest part of the job is getting the antenna in place. You'll need a few tools for the job; just which ones you'll need will depend on the antenna and how you're going to mount it. Typical of the sort of things you'll need are a hammer, a screwdriver, a crescent wrench, a roll of racer's tape, and a can of clear acrylic spray.

side of the dial; then switch to the ref mode and, with the mike still keyed, see where the needle falls. If it falls below 1.5:1 that's good; readjust the meter for channels 1 and 23, and if everything stays below 1.5:1 you're in business.

You can seek to optimize that reading, get it down as close to 1:1 as you can, by making very small adjustments of the antenna's tunable portion. You do it the same way that you would approach the problem if you can't get a 1.5:1 reading the first time around. Use an Allen wrench to loosen the Allen screw holding the tunable portion of the antenna. On top-loaded antennas this will be at the top of the antenna, and on bottom-loaded antennas this will be near the base. Make the antenna as short as you can, and go to the meter for a set of readings. If they're worse, start lengthening the antenna, in increments of about ⅛-inch, until the readings are what they should be, 1.5:1 or less on all three channels. If the readings are better but still not good enough, shorten the bottom end of the tunable portion of the antenna, just a little bit at a time, until the figures get right. Once you finally get the SWR right —and sometimes it takes a little time—you're ready to broadcast, providing you have your license. If you haven't, you at least can listen.

If you're about to install a base station things are a good deal less complicated. The major portion of your work is the installation of your antenna; that's just as time-consuming as a complete mobile installation, but you won't be messing about with power hookups and hole drilling.

The first thing you have to do is spot where you want your base station, and where you want your antenna. This is important because when you make these decisions you want to have an eye open to the requirements of keeping the transmission line as short as possible while mounting the antenna as high as possible. A certain amount of transmitter power is lost in the antenna cable, and the longer the cable, the greater the loss. You can gain back some of that loss, however, by mounting the antenna just as high as you can get it. For instance, a 40 percent increase in the height of your antenna will produce an increase in range of approximately 15 percent, while a 60 percent increase in height can produce a range increase of approximately 20 percent. So choose your locations carefully, and know exactly what you want to do before you start the job.

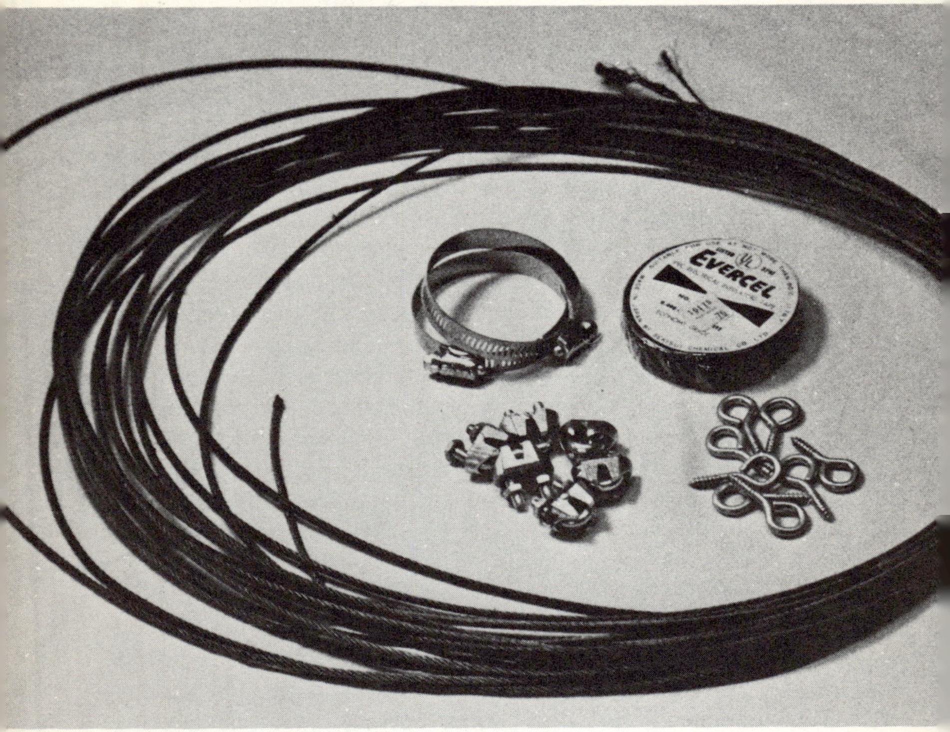

Figure 8-16. Along with the base radio, antenna, and transmission cable you'll need a few additional odds and ends, as pictured here. Among the things you might need will be a coil of wire to guy the antenna with, some eye screws, some Crosby clamps, a roll of electrician's tape, and maybe some pipe clamps.

Figure 8-17. First, lay all the antenna parts out somewhere and check to make certain they're all present and accounted for. Then read the directions through thoroughly.

Figure 8-18. Following the directions carefully, begin assembling the antenna. (The one you see in these pictures, by the way, is a Cushcraft Ringo.)

Tools for this installation, by the way, probably will include an eight-inch crescent wrench (known as a "metric adjustable" in some circles), regular and Phillips-head screwdrivers in medium size, a pair of diagonal cutters, a pair of regular pliers, a roll of racer's tape, and a can of clear acrylic spray.

Start off by opening the box your antenna came in and laying all the bits and pieces out. Read the instructions thoroughly so that you have a good understanding of what's required and make certain all the parts the instructions say you'll need are indeed part of the package. If they aren't, box the whole thing up and take it back to whoever sold it to you and ask for an explanation. If everything is there, go ahead with the assembly of the antenna. You'll want to carry this out in your garage or in your yard, or anywhere else that isn't on top of your house; no sense spending any more time up there than you absolutely have to. Follow the directions carefully and assemble the antenna *just* the way they tell you to.

Take the tools, your assembled antenna and the coaxial cable, the antenna mast, and whatever guy wire and fasteners you intend using up to the rooftop. Several ways to mount the mast exist, including tripods, clamps that mount on vent pipes, chimney mounts, peak mounts, and the like. Television-antenna hardware seems to work very well. Whichever sort of mount you choose, make sure the mast is fastened to something that is likely to stay put. Try to avoid mounting the antenna on a chimney, because the smoke and soot from the chimney can settle on the antenna and cause deposits that eventually will lower its efficiency. If you find what looks like an ideal mounting position that seems to be a bit loose in its moorings, give it a pass and look for the next most advantageous location. (An advantageous location, by the way, is one that keeps the antenna out of the way of tree limbs and the like that can hit it and brush against it during weather.)

Once you've found your spot, go ahead and fasten the antenna to the roof. Make sure that it is standing straight, and tighten everything down. If you live in an area known for its winds, you'll want to put up guy wires to help support the assembly. In most cases three of them will do the trick, placed at equal intervals around the base of the mast, with the wires or cables running at 45-degree angles or more. Put insulators into the guy wires and tie them into the mast below the antenna structure itself. If most of your wind comes from one specific direction, have one of the wires strongly mounted

Figure 8-19. When you get the antenna together, attach the mast to it. Do as much of this work as possible on the ground. The roof of your house is no place to spend any more time than is absolutely necessary.

Figure 8-20. Once you get the antenna together and the mast affixed, go to the top of your house, choose your mounting spot, and have someone hand you the antenna. If you choose to mount the antenna to a vent pipe, as our friend is doing here, make certain the pipe is firmly attached to the roof.

Figure 8-21. There are several ways of placing an antenna on a roof; television-antenna mounting hardware often fills the bill exactly. Here the antenna is being mounted to a vent pipe by means of a pair of pipe clamps. This method worked well under these specific circumstances, but might not have worked at all under different circumstances.

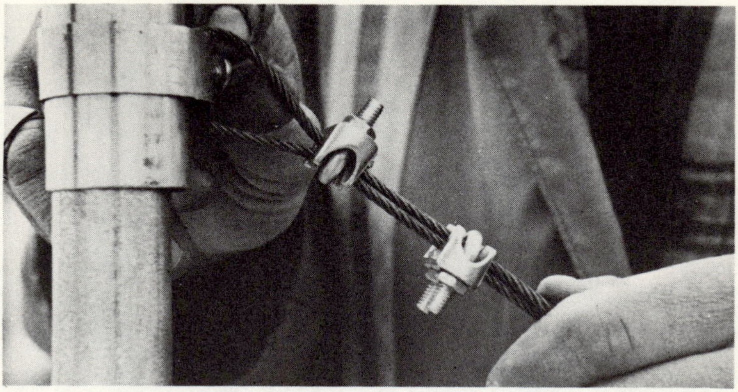

Figure 8-22. If you live in an area plagued by high winds, you may wish to guy the antenna. If so, use three wires placed at equal intervals around the antenna's base, and running at approximately 45 degrees. Affix one end of the guy cable or wire to the roof with an eye screw, then cover the screw with waterproofing compound. Run the other end of the cable to the antenna and use a pair of Crosby clamps to tie it down. Mount the clamps opposite each other, as seen here.

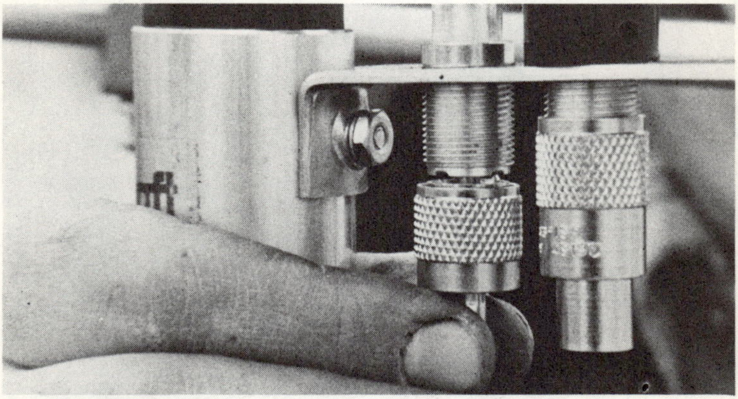

Figure 8-23. You're now ready to hook up the coaxial cable, which on this particular antenna goes right next to a short piece used as a tuning stub. When you hook the line up, make sure the teeth on the plug match the notches on the socket. Don't screw things down too tight, because you still have to check for SWR, and that'll mean loosening the cable once you've got the rest of the installation done.

Figure 8-24. Use a hunk of racer's tape to attach the cable to the antenna mast. Don't pull the cable tight; just tape it where it hangs, and tape it at the top of the mast and at the bottom as well.

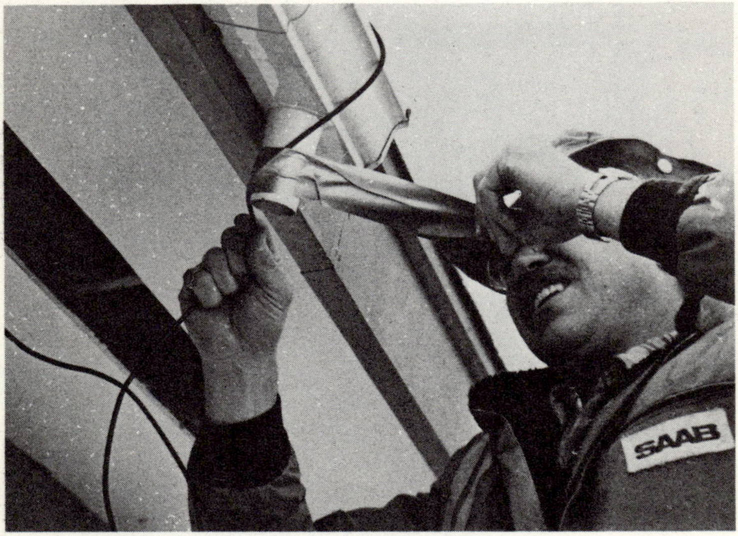

Figure 8-25. If you wish, you can let the cable lie on the rooftop. In this installation the cable was to enter the house through a roof vent, so it was taped to an adjacent downspout to keep it from flopping around in a stiff breeze

and arranged so that it will supply a direct pull on the antenna whenever the prevailing wind comes up. Now fasten the coaxial cable to the antenna and, to waterproof the coupling, spray the coupling with your can of acrylic spray. Run the cable down the mast and tape it securely every few feet, but don't pull it tight—leave just a bit of slack in it. Place the last piece of tape around the cable and mast right at the base of the mast and then run the cable toward the point where it will enter your house. You can let the cable lie flat on your rooftop or, if you wish, use standoff insulators.

Just how you get the cable into your house will depend a great deal on the house itself and on where you have the base station located, but there are at least three basic routes you can choose. You can run the transmission line into the attic via a roof vent, you can run it into the crawl space below your house via one of the bottom vents, or you can drill a hole in a wall just below a window frame and run the line in this way. The first two choices are probably more desirable than the last, because they cause the least amount of mess.

If you run the cable into the attic, you have to get it from there to the transceiver. One easy solution is to drill a hole in a closet ceiling and then run the wire down there, or to run it down through an attic crawl hole. If your house isn't built on a slab you can run the cable under the house, drill a hole in the floor in an appropriate location, and feed your cable in that way. If neither of these approaches appeals to you, you can drill a hole in a wall and run the transmission line in through there. Be sure you caulk the hole up good, or outside weather will get inside.

Whichever way you've chosen, you've now gotten the transmission line in to your transceiver. But you'll need to guard against lightning, so before you screw the cable into the back of your transceiver, attach a lightning arrester to the cable's end. Then run a 12- or 14-gauge copper wire from the arrester to a cold-water pipe and thus ground the antenna.

Another potential problem you can head off right now is television interference (TVI). This is a situation in which random interference results from your CB broadcasting and plays hell with the pictures on the television sets belonging to your neighbors. To avoid having a lynch mob come after you later, let your neighbors know what you're doing now, and give some strong consideration to installing something called a low-pass filter in your antenna transmission line. It is interposed in the

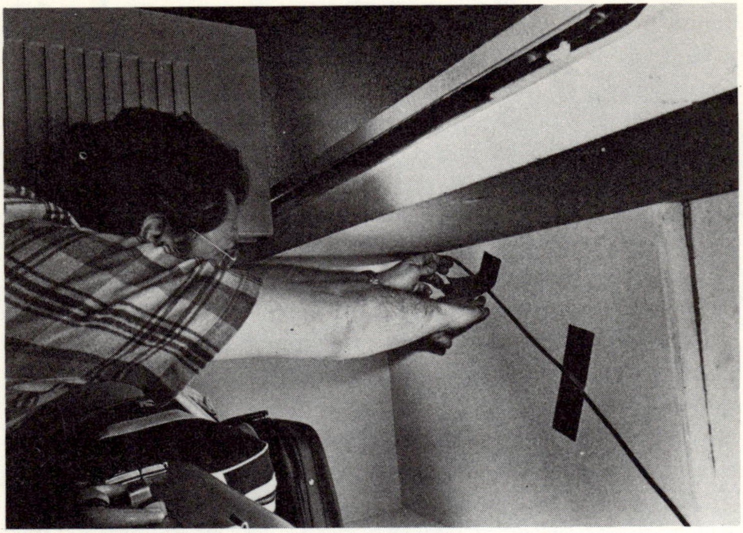

Figure 8-26. Once inside the attic, the cable in this installation proceeded to a bedroom closet, where it entered the room via an attic crawl hole. To keep the installation as clean as possible, the cable was taped to the closet ceiling.

Figure 8-27. You of course know before you start where you want the radio to live, and now is the time to put it there. (The radio in this installation, by the way, is a Midland Model 13-863 AM set.)

Figure 8-28. This is a lightning arrester, and it's a good idea to have one interposed in the transmission line between your set and the antenna. If you don't have one, a bolt of lightning striking your antenna could ruin your equipment and start a fire. The other end of the solid copper wire that you screwed to the ground clamp gets screwed to the fixture on the side of the lightning arrester.

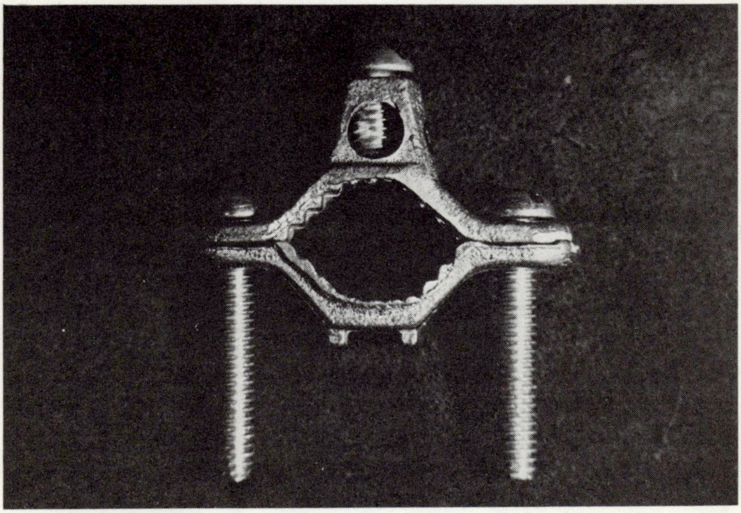

Figure 8-29. This is a ground clamp. You bolt it to a cold-water pipe and then screw a solid-core copper wire to it to provide a ground for your lightning arrester.

Figure 8-30. The arrester itself screws to the cable socket on the back of the radio.

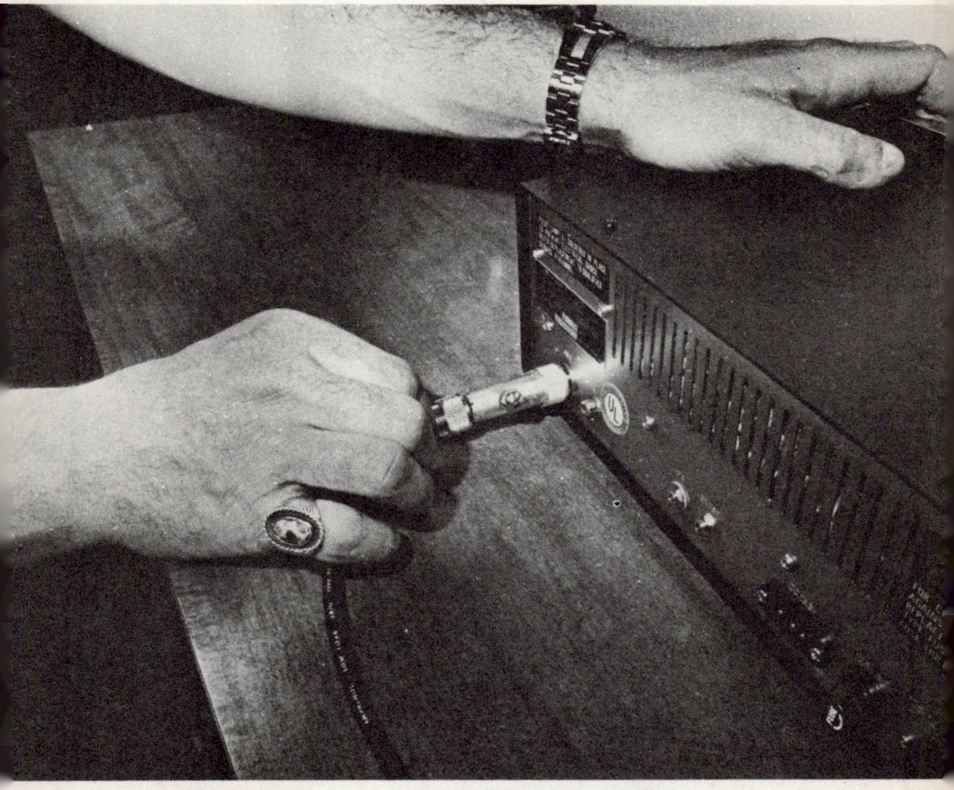

Figure 8-31. That done, the cable merely screws to the arrester. A TVI low-pass filter works the same way.

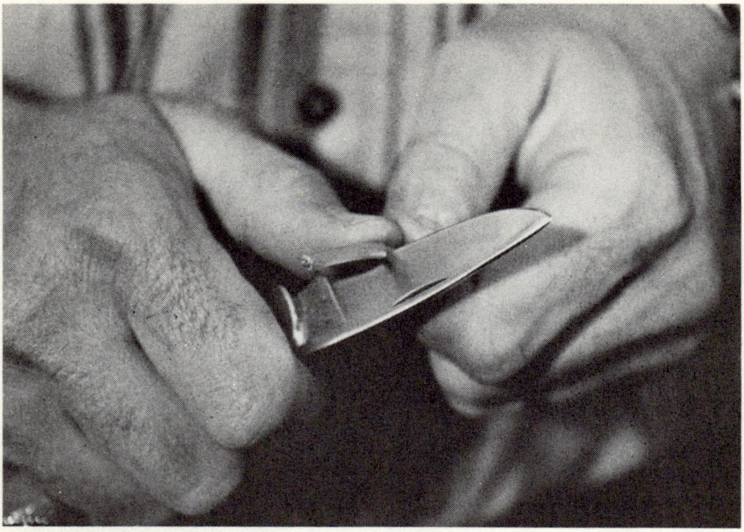

Figure 8-32. The arrester in place, you're ready for your copper wire, either 12- or 14-gauge. Strip about an inch of insulation from the end of the wire with your pocket knife.

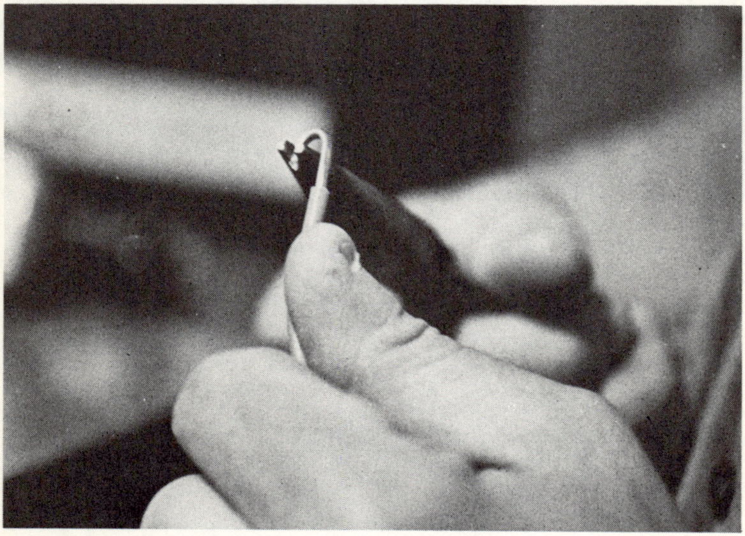

Figure 8-33. With the insulation off, take a pair of needle-nose pliers and bend the end of the wire into a hook. Then screw it to the side of the lightning arrester.

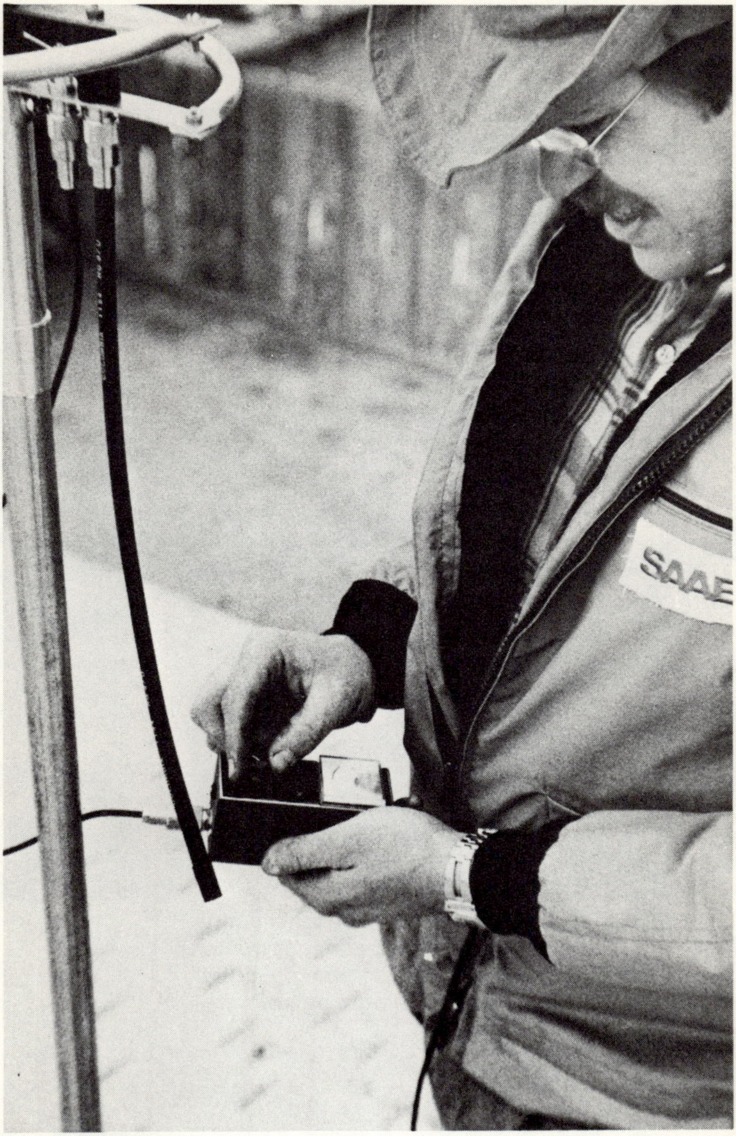

Figure 8-34. Now go back to the rooftop and hook the SWR meter into the system there, *not* right at the radio. Make certain the SWR readings are no more than 1.5:1 on channels 1, 11, and 23.

Figure 8-35. This particular antenna uses a tuning ring to adjust the SWR, and that's what our friend is adjusting here. The ring is locked into adjustment with a screw bracket, seen here right by our friend's hand.

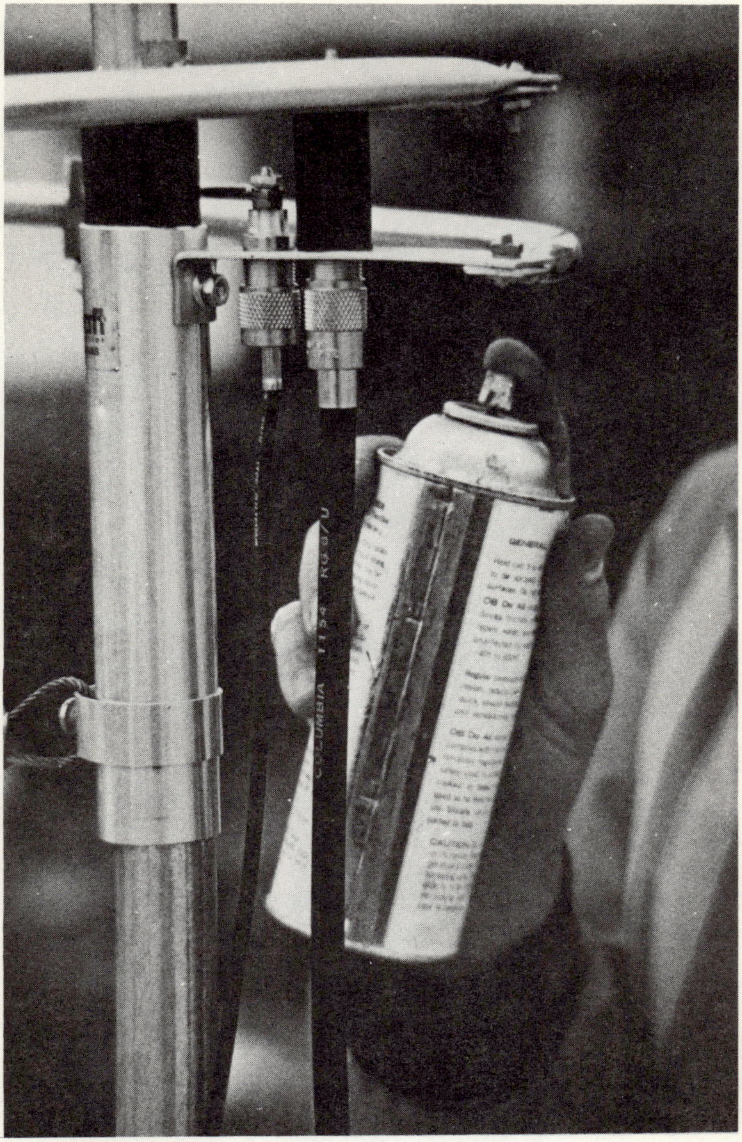

Figure 8-36. When the SWR is what it should be, remove the meter from the system, replace the cable on the antenna properly, tighten the coaxial plug's collar down firmly, and spray it with clear acrylic to keep the moisture out.

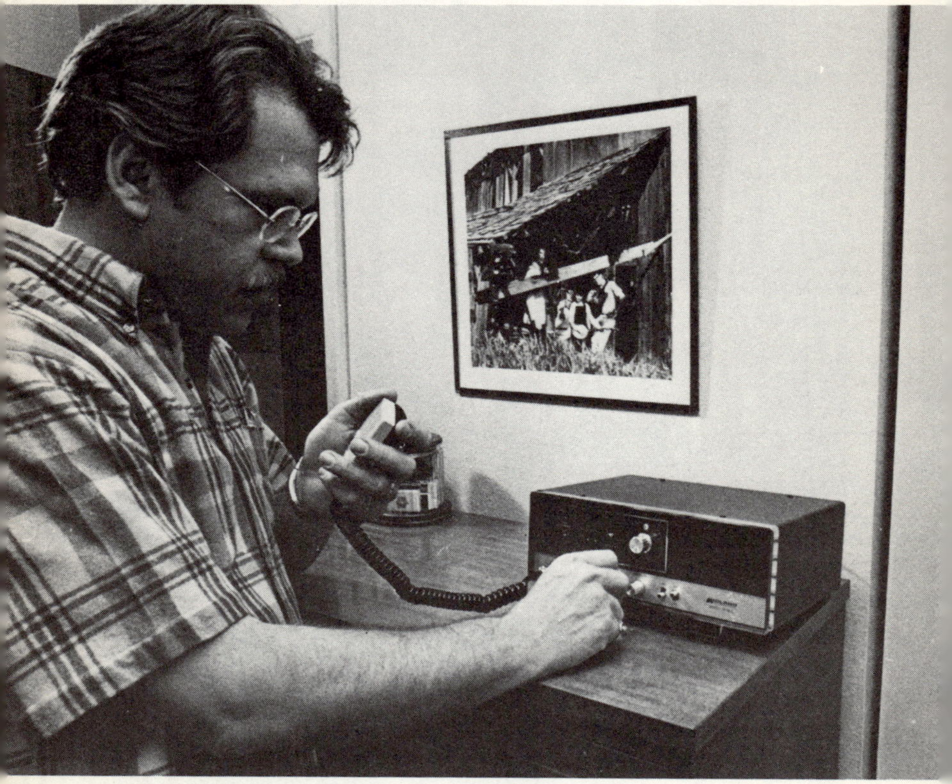

Figure 8-37. You now have a complete installation, and once your license arrives in the mail you'll be able to broadcast.

cable just as the lightning arrester is, and effectively puts an end to most television interference.

Before you operate your base station you'll want to check the SWR. Follow the same steps mentioned earlier for checking the SWR of a mobile installation. Base station antennas have different kinds of SWR adjustments on them than do mobile whips, so do what the antenna directions tell you. That done, and CB license in hand, you're ready to broadcast. And once you start, you'll find that with both base station and mobile use, you've got a little electrical interference. There are two kinds of interference, natural (atmospheric electrical noise) and man-made (the kind that comes from things like automobiles, electrical appliances, and some types of lighting equipment). Noise gets into a radio system through radiation sensed by the radio's circuitry through its power cord antenna. You'll mostly hear natural, atmospheric noise on a base station, and there isn't too much you can do about that. You may also hear a small amount of noise created by nearby electrical appliances, and this you can beat by installing a line filter in your station's power cord.

You'll hear mostly man-made noise on your mobile set, and there is a great deal you can do about that. It's picked up from two main sources: the car's ignition system and its electrical system. Ignition noise has a popping sound that has no pitch and increases in frequency as the speed of your car's engine increases. Most of this can be eliminated by using noise suppressors between the spark plugs and the spark plug wires, by using noise-suppressing spark plugs or radio-resistant spark plug and ignition wires, or by putting a noise suppressor on the main coil wire.

The other kind of noise you're likely to hear from your mobile unit is a high-pitched whine that increases in pitch as the speed of your car's engine increases. This comes from the car's electrical system. If your car has a generator, that's probably the main source; if it's equipped with an alternator you're in good shape, because these tend to make much less noise than do generators. The noise produced by generators comes from the electrical energy arcing between the generator brushes and the armature. There's a cure. It's called an inductive coaxial, or bypass, capacitor, and it is mounted securely to the generator body. It's self-grounding and needs a good clean contact, so before you fasten it down clean a spot on the generator and the foot of the capacitor with steel wool to get a little bare metal showing. You may find that a standard capacitor already lives on the flank of the car's generator. If so, it has been designed to keep electrical noise out of regular AM reception, and it won't be much use for Citizens Band radio reception. Get rid of it and put the new unit in its place. When you do the hookup, make certain that you get the wires to the armature lead, not to the field lead. If you do hook up to the field lead and the capacitor shorts out, you surely will have to buy a fresh generator, and that's expensive.

While we're talking about your car's electrical system we should note that, while the average automobile electrical system has heavy work to do, solid-state Citizens Band radios make a relatively small power demand. If after hooking your radio up you have trouble keeping the system charged, or if an inductive coaxial capacitor does nothing for your electrical-noise problem, chances are that you have a problem somewhere in your car's electrical system and should have it looked after. That may sound like a bummer, but it's far better to find out about any potential problems in this manner than to have the whole system surprise you by going dead some dark night just when you need it the most.

One last kind of noise you're likely to encounter is from static electricity. You'll recognize it as a popping noise that comes from your radio's speaker when your car is in motion. A prime cause is looseness in body and frame parts. This is a problem more likely to be found in older cars than in newer ones, and it can be fixed by making sure everything is good and tight. Some static-electricity interference also can be caused by the rolling motion of the car's wheels, and this usually can be fixed by installing wheel-static-eliminator springs in the grease cups of the front wheels.

All right—if you've paid attention to the instructions that came along with your radio and antenna, and if you've followed the guidelines given you in this chapter, chances are that you've got a nice clean radio installation, one that you can be proud of, and one that will live through high winds, fast highway speeds, and extended use. And that's exactly the idea. To get the most out of your installation, though, you'll want to find out all you can about running your station, learning the rules of the road, and following correct broadcast procedures. So read on.

9

OPERATING A CB STATION:

How to Make It Work for You

If you've followed along this far, you should be just about ready to begin operating on Citizens Band radio. You've taken considerable time and care to educate yourself about the hows and whys of CB radio; you've taken an equal amount of time and care to choose the radio set you think will best suit your requirements; you've applied for a Citizens Band license and you have it in your possession; you've installed your radio and your antenna so that the job is neat and professional in appearance and sturdy enough so that you'll not have to do it a second time—and you're ready to begin.

The first step? That should be to decide for yourself here and now that you always will conform to the spirit and letter of the Federal Communications Commission's rules and regulations whenever you operate your set. So sit down right now and read through Part 95. If you're going to conform to the spirit of the FCC's rules, that means you'll need to have a copy of your license handy, so whip on down to your local library, or anywhere else that has a coin-operated photocopier and run off several copies of your license. If you've got a mobile unit, stick one of the copies in the glove compartment of your car or truck. Put it with your other important documents—registration papers and the like—so you'll always know where to look for it. Then take a transmitter-identification sticker (remember the little tag you asked the FCC for when you wrote requesting your copy of Part 95?) and write all the necessary information on the sticker. Peel off its gummed back and place the sticker on your transmitter in a spot where it can be seen. That done, get yourself a small loose-leaf notebook to use as an informal logbook. You probably won't be making frequent entries in it, since it's not all that wise to drive, use your radio, and write in a little book all at the same time, but you can use the book as a place in which to maintain necessary records. Find a copy of the 10-code that you want to use and fold it in between the pages of your book. On the first page, record the date you installed your equipment, and in what sort of vehicle. Write down the serial number of your radio unit to aid in its recovery if it should ever be stolen (and thefts of mobile Citizens Band radio units are rising rapidly. If CB is the hot new item for the electronics industry, it also is the hot new item for every ripoff artist in town).

When you adjust your set's SWR, use your book to record the date, the location and the lowest readings you settle on for channels 11, 1, and 23. Each time you recheck the SWR make a fresh entry, thus keeping track of just how much your set's SWR changes. When you have trouble with the radio, or when you have any kind of servicing done, make an entry in your logbook, and if you should ever add accessory equipment to your set, record that in your book as well. If you plan on monitoring channel 9 you'll probably want to record the call numbers of the people contacting you and the time of the call.

If you have a base station you're going to want to do some of the same things. Get yourself an inexpensive picture frame and hang a copy of your license right near your radio. Put the original in your safe-deposit box along with your other important papers—the deed to your home, insurance papers, business documents, and the like.

And, just as you would if you had a mobile unit, get your hands on a transmitter-identification sticker and, after filling in all the appropriate information, stick the thing to the body of your radio set where it clearly can be seen. Now you're official.

If a logbook is a necessity for a mobile unit, it's even more of a necessity for a base station, and in the long run could be a whole lot more valuable. Unlike the guy who is driving along using his mobile unit and trying to keep his equipment on the road, you'll be standing or sitting in front of your radio, surrounded by all the comforts of home or office, and you'll be completely able to make complete entries. So get yourself a full-size loose-leaf notebook. Use it as a logbook and as a binder for Part 95 updates and revisions as they arrive. On the first page of the book write your name and address, the brand names and model numbers of your radio and antenna, and any serial numbers that are appropriate. On the second page record the date you hooked your equipment up and all the SWR information you generated the first time you checked the standing wave ratio. Each time you check your equipment in the future—and you'll want to do that every few months—add all the new SWR information.

Just how thoroughly you fill out the logbook is up to you. Some operators will refer to it but seldom, making an entry only when some sort of servicing is done to the transceiver or antenna, writing in just what was done, the date, the time, the cost, if any, and by whom the servicing was done. Others keep right on top of their books, recording in addition to the above every transmission made over the radio, its date and time, weather conditions, who the conversation was with, its subject, and whatever other details the operator feels are necessary. For the time being, try to aim for a middle-of-the-road course. You don't need to record every last detail of every call, but if, for instance, you should call one of the service organizations working channel 9 to ask for a weather report for a nearby mountain pass, or to report a nonfunctioning traffic light, record the date, time, and whatever the unit you reach tells you. A logbook can be a handy reference document—but only if you decide going in that you're going to do it right. Don't forget that you should have a copy of your favorite 10-code either written somewhere in your logbook or tucked between the pages where you can find it fast when you need it. And keep a reliable ballpoint pen or a pencil handy so that you always have something with which to write.

Something else you're going to want to be certain you have right by your radio, logbook, and pencil is a reliable clock. If you wear a watch, that's fine, and if you bought a base station with a timepiece built into it, that'll do also. If neither of the above is true, go and buy yourself an inexpensive electric clock with a lighted dial that you can trust to keep reliable time. It'll help you keep close track of when you make your calls and receive them, and it'll be handy when some other CBer calls and asks you for a 10-36 (time).

A couple of electric appliances will come in handy as well. The first is some sort of electric light. If your radio is located so that the main room light will illuminate it, you're in luck, but if you have your radio somewhere where the light is faint, or in a bedroom where a strong room light might disturb someone, buy yourself a small desk lamp and place it next to the radio so that you can see what you're doing. The other item that will come in handy is a telephone. Probably you won't be obliged to use it often, but from time to time you'll hear CBers calling for someone with access to a land line (a telephone, in CB lingo). The caller might be out of gas, he may have a flat tire, he may be stopped at a local saloon, and he may want you to call for help or notify his wife. If you choose to answer his call, your telephone will be right there and you won't have to leave your radio to use it.

So if you've got your radio in operating condition—your license posted, your transmitter-identification sticker in place, your logbook and pencil at hand, a clock nearby, enough light to see by, and a telephone handy—you're ready to go. Make sure everything is plugged in and the set's squelch is turned all the way down, all its noise-limiting equipment is turned off, and the channel selector is tuned to channel 1. Turn on the main power switch, and as soon as you begin hearing radio transmissions, adjust the volume control. Starting with channel 1, tune to all twenty-three channels and listen briefly to what's going on out there so you can get the lay of the land. Although channel 11 has been designated the national call or hailing channel, not all locales use it. If you hear nothing happening on 11, find yourself a channel that does have some action on it and settle in for a bit. The object here is to make your first call, and to do that you have to find a channel with someone on it besides you. When you've found what sounds like a workable channel and you've listened in for a bit, begin adjusting the radio's squelch knob. Turn it up just far enough to cut off any between-transmission noises, but not so far that it cuts off everything but the very strongest signals. Keep adjusting it from time to time and develop your touch. There's no point in having the squelch setting any higher than it absolutely has to be, so find out just how far you have to go. If, for instance, you can hear one side of a conversation but can't hear the other side, you may have the squelch turned up too high—or you simply may be out of range of the transmission you can't hear. The only way to find out is to diddle with the squelch adjustment until you have a full understanding of what it can do for you.

Once you're got your radio's squelch knob handled, keep listening, and begin turning on your noise-limiting switches. Turn them on, listen; turn them off, listen. If it looks as though you don't need them, don't use them. If you're running a base station, chances are that from time to time you can get along just fine without them—and perhaps hear really weak signals that the noise-limiting circuits in your set could interpret as noise and block out. If you're in a mobile unit, chances are that you'll want to use all your set's noise-limiting and noise-blanking capacities, particularly when your car's engine is running. You should be aware that if you're in a mobile and you're having trouble receiving a distant station, there are some steps you can take to pull that signal in. Chief among them is stopping your car and shutting down everything but the power supply to your radio. If you can, get close to the top of a hill. Remember, antenna altitude relates directly to Citizens Band radio range.

Now that you've got your squelch set and have given some thought to your need for noise limiters, it's time to make a call. Its purpose will be to see how well your set is working, and to initiate it you wait for a break in the action on your local call channel and say, "Break." (If you've got a hand-held mike, by the way, remember to hold it about four inches away from your mouth, off to the side just a little and at a slight angle, so that any breath escaping between your lips as you speak won't jar the microphone's sensitive element and cause a popping sound. If you've got a desk-top mike, keep it about eighteen inches away from you and speak directly into it.) After you've said, "Break," someone will come back over the air and say something like "Go ahead, Breaker," and when you've gotten this go-ahead, give your call sign—ZZZ5555, say—and ask for a radio check. This would sound more or less like this: "This is ZZZ5555, the Los Angeles Bear Buster, thanks for the break, and can someone give me a radio check? Over." Chances are someone out there will be happy to work with you, and if you do find someone, turn to another channel (call channels are for calling, right?) and talk to him for just a bit to find out how well he's receiving your signal and how well you're receiving his. Your radio's S meter will be a help here, because it will show you the relative strength of his incoming signal, which you can relay back to him. And if his set has an S meter, he'll be relaying what it tells him back to you. Write the figures, and the location he's broadcasting from, into your logbook.

When you've found that your station is operating clearly and efficiently in both the send and receive modes, thank your new friend and sign off by saying something like "Thanks for the radio check, this is ZZZ5555, we're clear." Simple, yes?

You've just opened a whole new world and given yourself access to information that can be gotten in other ways, but nowhere nearly as quickly. You want to know what the weather is like over the infamous Donner Pass or on the road to your favorite resort? Call for a local REACT unit on channel 9 when you arrive in the area and ask. Lost in a strange town and need directions? Go to a call channel and ask for help. Out on the highway wondering where the highway patrol is and what it is doing? Hail the first eighteen-wheeler you see that is wearing a CB antenna and strike up a conversation; the driver probably will be able to answer your question. Before you've had it long you'll find many uses for your set, and if you're like most folks, you'll wonder what took you so long to get into CB.

But if CB is infinitely useful, it also can be a tremendous pain in the neck. You'll find out just how big a pain when you begin monitoring your area's regular call channel. These sometimes are known as "garbage channels," and it won't take you too long to figure out why: You'll hear everything from the random ramblings of drunks to the uninhibited patter of small children. You'll hear teen-agers trying to set up dates and grown men spoiling for a fight. You'll hear people attempting to regulate a channel as if it were their own private property, and you'll hear others sending out what are called dead carriers—keying a CB radio set's microphone so that the transmitter section sends out an unmodulated carrier frequency that disrupts normal communications on the channel. That's all right; let those others abuse the service. The FCC will catch up to them soon enough if they keep at it. But in the meantime, when you find such craziness going on, switch over to channel 9. If you can't count on finding sanity anywhere else, you usually can find it there—but even that isn't guaranteed. As you already know, the FCC has designated channel 9 as the national CB emergency channel, and usually it is monitored by a member of one of several Citizens Band radio-service groups. Two of these groups that are particularly well known are

REACT, 111 E. Wacker Drive, Chicago, IL 60606, and ALERT, National Press Building, Washington, DC 20004. If you have any questions about their activities or want to be involved in what they're doing, write. If not, know that by listening to channel 9 you have a pretty good chance of knowing all about traffic tie-ups, accidents, and changing weather conditions that could affect driving safety as soon as they become evident. When you're out on a long drive you'll find it useful to scout around until you find the channels the truckers are using, and switch over to 9 as you go through cities and towns. The truckers will keep you advised of what is going on out on the highway, and the good folks of REACT or ALERT will keep you informed about what is happening in civilization.

Incidentally, while we're on the subject of channel 9 and its function as an emergency channel, you should know that the FCC has had some very specific points to make about emergency calls. FCC Rules and Regulations Part 95.41 says, "The following priorities should be observed in the use of channel 9. 1. Communications relating to an existing situation dangerous to life or property, i.e., fire, automobile accident. 2. Communications relating to a potentially hazardous situation, i.e., car stalled in a dangerous place, lost child, boat out of gas. 3. Road assistance to a disabled vehicle on a highway or street. 4. Road and street directions." And that's not all. Part 95.85 gives some additional guidelines for emergency communications that you should be aware of. It says, "All citizens radio stations shall give priority to the emergency communications of other stations which involve the immediate safety of life of individuals or the immediate protection of property." These two clauses alone, which point out Citizens Band radio's readiness to be of assistance in the event of an emergency, are a big part of the reason why so many people now travel with Citizens Band radio in their cars.

Part 95 gives some other guidelines for the use of your radio, and you should be completely familiar with them so that you can get the most of what CB radio can offer you. Mostly they involve rules of the road, daily operating procedure. Follow them and you'll be doing it right. Don't follow them and you'll be looking for a call from a representative of the Federal Communications Commission. In a nutshell, they're as follows: You must broadcast your call letters at the beginning and at the end of each transmission. You also can use your handle if you wish, any phony name you've dreamed up to give you a radio identity, but you must always use your call letters first and foremost. With the exception of channels 9 and 11, you can call on any station you wish, and you can talk for up to five minutes. Keep track, and once you've been on the air for five minutes, sign off and stay off for at least sixty seconds. There's no point in monopolizing the frequency. Use nothing with the radio that illegally will modify its performance. This means that microphones that boost your set's modulation beyond 100 percent and linear amplifiers that boost your transmitter's signal beyond 4 watts should be completely avoided. Not only will the FCC crack down on you if it so much as finds a linear on your property, hooked up or not, but use of equipment like this can greatly deteriorate the usefulness of all twenty-three channels, not just the one you're on. Putting too much power out causes your signal to bleed, or slop over, into other channels, sometimes all of them, thus totally disrupting normal communications. You've got nothing at all to gain by getting every other CBer in town out looking for you with blood in his eye.

Naturally, it is illegal to use Citizens Band radio for any purpose that is against the laws of the federal, state, or local government—neither is it very wise, since anyone at all can listen in on CB conversations, and all kinds of people do. You are prohibited from using the word "Mayday!" unless there is grave and imminent danger to life and property; neither can you use any other international distress signal casually. And, last, remember that the airwaves used by Citizens Band radio communications belong to everyone—they're public property —and for that reason you must refrain from using indecent language, communicating obscene messages, whistling, singing, and sending music or audio tones. In other words, when you're on the air you must act like a proper, civilized adult who has a natural respect for the rights and feelings of other people.

And, finally, you should know that if you travel with your car into either Canada or Mexico, you can use your CB radio there as long as you have had the foresight to obtain the proper written permission. Find out about it when you make your travel plans.

These are but the broadest of guidelines, and you'll learn more as you go along. You'll also learn that there are a whole lot of CBers out there who have taken none of the steps you have taken to prepare themselves for proper use, understanding, and appreciation of this singularly useful form of communications. Too bad. They can sure enough make things inconvenient for you, but mostly the loss is theirs. What information you've picked up as you've gotten ready to broadcast on CB radio will be useful to you for many years to come—because you've gone at it systematically. You now have a rough idea of what makes radio work, you know the rules of the road, and you can walk into any store anywhere and pick a good radio out of a covey of bad ones. All you need to know to make you the Compleat CBer is a little information about the various accessories that are available for use with Citizens Band radio.

10

CB ACCESSORIES:

How Many Do You Need?

Citizens Band radio is no different from any other product that you can buy off a store's shelf. It's been designed to do a certain job, and in most cases it will do it quite well, but, just as there are certain added items you might want when you purchase a camera, a weapon, a stereo set, or even a car, there are certain items that can improve the performance of a Citizens Band radio or lengthen its life, or both. If you so choose, you'll be able to get along quite well without most of these items, because, if for no other reason, once you've got your radio and antenna set up, you've got yourself a complete, working unit, and many add-ons are frills that you may or may not need. Some of them, however, are bona fide boons to your radio's reception, reliability, and convenience, and you should know about them.

A lightning arrester, for instance, is something you should give some serious consideration to if you've just installed a base station in your home or office. Because if you've made such an installation, you've stuck an antenna on your roof that just could get struck in a storm. If a good healthy jolt of lightning struck your antenna, it would feed prodigious amounts of energy into the antenna, down through the coaxial cable, and into the transmitter. The antenna likely would be a total loss, the cable would be cooked to within an inch of its life, and there is a chance that your transceiver's innards also would be ruined. A lightning arrester plugs into the transmission line just before that line plugs into the transceiver. To ground the antenna, you run a copper wire from the screw terminal on its side to a ground. You need to keep the wire as short as possible, so if your house has wood floors, use a small drill bit and an electric drill to poke a small hole in the floor right by the transceiver and run the wire down through it. You will in the meantime have made a trip to a hardware store to buy a small piece called a ground clamp. Take the clamp, and with your crescent wrench, crawl under your house, bolt the clamp to a cold-water pipe, and attach the ground wire to it.

If your house or office is built on a slab, run your ground wire outside by the shortest way possible and tie it into a copper rod that you've driven at least five feet into the ground. Then, if lightning should strike your setup, you'll still probably sustain some damage, but chances are your equipment won't have to be completely written off.

Another accessory you may wish to consider is an auxiliary antenna. This is the sort of thing that wouldn't be useful for everyone, but suppose your base station is built around a good omnidirectional antenna but that arrangement doesn't give you quite enough range to talk to other stations on the far side of your town, or to an outlying REACT station, or to the shop of a business associate. You might not wish to give up the convenience of your omnidirectional antenna by taking it down and replacing it with a beam, so what you could do is mount a second antenna on your roof. This second antenna probably would be a good beam antenna, one with a considerable amount of gain.

If at some future time you decide to go with a pair of antennas, in addition to the antenna itself and a batch of coaxial cable you'll need a coaxial switch. It will have provisions for plugging in three different transmission lines. Into two of these receptacles you'll attach the leads from each of your antennas, and into the third you'll attach a piece of transmission line that will connect the switch to the transceiver. More than likely, the coaxial switch will have a sliding switch to allow you to choose which antenna will get the benefit of the transceiver's attentions. To use your omnidirectional antenna, the switch would be in one position; if you were to switch to your beam antenna, the switch would be in the other position.

If one transceiver can use two antennas, can two trans-

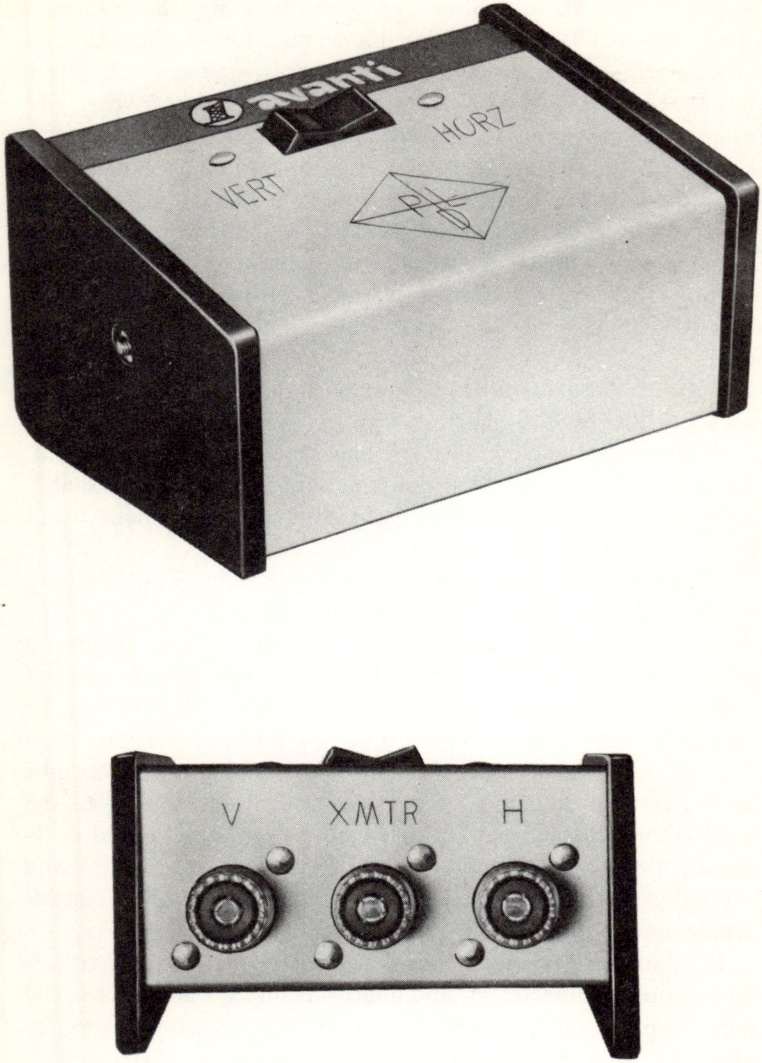

Figure 10-1. Sometimes necessity dictates the use of two different antennas and this box, called an antenna switcher or coaxial switch, is what makes it possible. A cable from each of the antennas is plugged into the box and a third cable, going to the transceiver, also is plugged into it. A switch on top of the box then decides which of the antennas will be in use. *Photo courtesy of Avanti Research & Development, Inc.*

Figure 10-2. Power converters come in several shapes and sizes, but they all do essentially the same thing — allow you to run a set built for 12 volts DC on 110 volts AC. *Photo courtesy of Breaker Electronics.*

ceivers use one antenna? Most certainly, but it takes a unit called an antenna coupler to make this work. An antenna coupler looks very much like a coaxial switch in that it too has provision for attaching three different coaxial lines and a sliding switch. What it does is interconnect a pair of transceivers and a single antenna, and it is the sort of thing that might be useful if you wanted to keep one transceiver in your home and another one in a nearby shop or barn. It would allow you to avoid the expense and hassle of buying and installing a second antenna by letting both receivers listen at the same time. Just one of your transceivers would be able to broadcast at a time, though, and you choose which set will do the work by throwing the switch on the antenna coupler to activate either transmission line.

If you need a base station but you're on a budget, or if you wish to make your mobile unit do double duty, there is a way to arrange a typical mobile radio so that it will perform like a base unit. What you need for this is a power converter, an electrical device that converts the 110-volt alternating current that flows from the wall sockets in your home to 13.8 volts (i.e., nominal 12 volts) of direct current, the kind of power that flows from your car's battery. By obtaining one of these to go with your mobile unit, you can get into a base station for a few dollars less than it would cost you to get into a standard base unit, and that's a good advantage. But there are disadvantages as well. The first and most apparent of them is that if you intend using your radio both in your car and on the power converter, you will have to go through the hassle of mounting the radio before you can bring it inside and hook it to the converter. You can beat this by buying a quick-release radio mounting bracket and quick-disconnect adaptors for your power and ground lines. Your CB store should be able to obtain both these items for you. The second disadvantage is that there are power supplies on the market that will make the radio work but won't really give you the kind of performance that either a good base or a good mobile unit, properly installed, will provide. A specific problem found in some of these converters is that when you turn one of them on and begin pulling power from it, the voltage can drop and, with it, the amount of power sent to your antenna by your transceiver. Another problem is that many radios these days are built to adapt to cars that are either positive or negative ground; the radio itself is built to adopt the polarity that the car's dashboard runs on. This means that the radio has what's called a floating chassis; it is neither positive nor negative. If a radio of this sort is used with a power supply, chances are it will remain totally insulated from it; the only connection will be the 12-volt dc wires between the two units. With this sort of an arrangement it is possible that any noise radiated from the power supply could be picked up by the radio's circuits and reproduced as an annoying hum. You don't need that. There also is the possibility that a radio and a power converter may be of different ground potentials, and if this is the case you just could learn about a phenomenon called ground loop. Ground loop is a condition in which the power current in a single ground wire can cause electrical instability—and possibly a slight shock when you contact the transceiver/converter unit.

Heard enough to be completely turned off on this power-converter idea? Well, don't be, because, as we said before, this is not necessarily an unworkable compromise. You *can* buy a power converter that works well, provides a good, even flow of power, doesn't make noise, and doesn't jolt the stuffings out of you when you touch it. You just have to know what to look for. For starters, look for a unit with a very well filtered and regulated power supply. Look for a unit that will produce a regulated 12.5- to 13.8-volt dc shot of current, and one that is capable of handling 3 to 5 amps of power. Better yet, if the manufacturer of the mobile unit you've chosen makes a power converter, chances are it will have the right specs and work beautifully with your set. Check those specs out and then buy it if they're what they should be. That just may be the safest way to stay out of trouble with a power-converter unit. Most of all, don't be put off by the drawbacks we've mentioned in connection with power converters. This is as good a way as any to get a workable, reliable radio at budget prices, and if space is going to be a problem with your planned base installation, this sort of an arrangement could help out. Just do it right, and don't go off half-cocked and buy a cheapo unit.

A device that is not absolutely necessary but that many CBers wouldn't want to be without is an SWR meter. We dealt with what this little jewel is and what it does pretty thoroughly in the chapter on installations, and every several months or so you'll want to make a check on your set's SWR to make certain that your system is properly tuned and working at maximum performance. To do this you're going to have to borrow a meter somewhere, plead with the man at your local CB store to help you out, or own a meter of your own. They're not expensive—basic units can be had for approximately twelve dollars—and if any accessory should be considered a necessity, perhaps it should be this one. As mentioned earlier, these meters let you see just how well your radio, coaxial cable, and antenna are matched, and they let you make whatever adjustment you need to correct a mismatch. But there's another side of the coin. Built into many SWR meters is a device known as a field-strength indicator. To make it work you screw a small stub of an antenna into the meter box and, while your radio is transmitting, take a reading from the power radiated by the antenna. Once you've established what a normal power reading should be, you can quickly see when a problem comes along to cause your set's power to drop off. The meter won't tell you what's wrong, but it will let you know that all is not as it should be, and that might be reason enough for owning one.

Suppose, just suppose, that you get the antenna to either your mobile unit or your base station hooked up, you whip your SWR meter onto the set—and you can't get a reasonable SWR reading to save your fanny. You've got a problem, and

Figure 10-3. If, once you get your radio and antenna installation done, you have trouble getting the correct SWR reading, what you may need is one of these: an antenna matcher. It is interposed in the transmission line. The knobs on top of the box allow you to make the necessary SWR adjustments. *Photo courtesy of Avanti Research & Development, Inc.*

what you may need to fix it is a device known as an antenna matcher. It is interposed in the antenna line and can be adjusted to match the output impedance of the transceiver, antenna, and transmission line. Some of these are weatherproof and are intended for outdoor use; some are not and are meant to be kept inside. Some of them have SWR meters built into them; some don't. Don't buy one of these unless you need it, but know that if you have trouble and do need it, it's available.

One accessory rapidly growing in popularity is the auxiliary speaker, an add-on speaker that is connected to your mobile unit's audio circuits and mounted in the car away from the radio. Sound like an unneeded trinket? Well, maybe, but there is a good reason for this sort of thing. Most radios are just radios. Their designers have taken great pains to make them do what they do with the utmost efficiency while consuming the smallest possible amount of space, and as far as audio reproduction is concerned, that lack of space is the problem. Once all the radio circuitry is shoehorned into its case, there usually isn't a whole lot of room left over in which to put a speaker, unless it's a very small one; neither is there a great deal of room in which to install the baffling necessary to make the speaker sound as it should. As a result, many radios are delivered with truly tiny speakers, and while most work just the way they're supposed to work, often the quality of the sound they reproduce isn't all that great.

An auxiliary speaker, on the other hand, usually has been designed to give the best possible reproduction of the human voice, not to fit into the smallest possible space. It's not just the speaker that does it, as noted; the baffle box plays a large part in how true the reproduction is. Certain kinds of baffle/speaker arrangements work best in certain kinds of audio ranges, and those sold as CB accessory speakers for the most part have been designed specifically to make voice radio communications more intelligible. A good one will take out all the tinny treble sounds and all the bass sounds you may be used to hearing and leave you with improved voice response.

There are two different types of speakers, and you usually can find them at CB stores or at stores that specialize in ham radio equipment. The first is the all-weather variety that you mount out in the weather, maybe behind your car's grill or on a roof rack. This lets you monitor your radio when you're out of the car, and if your set has a public-address function you could use it for that as well. The other kind of speaker has been designed to replace the one in your radio, and it usually has been designed to live indoors, or, at least, inside, out of the elements. In most cases it is a small plastic box, maybe five inches or so square and two or three inches deep, and comes with its own mounting bracket. It also comes with a cord and a plug, and many radios have jacks into which you can insert the plug and immediately be in business. Some radios have no such jack, but the man at the CB store can probably adapt your radio so that it will do what you want it to do. If your radio has a jack, so much the better. If you can, before you

Figure 10-4. There are many different types of accessory microphones. This one, designed for use with either a base or mobile unit, features a built-in volume control, the aim of which is to fully modulate a transmitter. *Photo courtesy of Turner Division, Conrac Corporation.*

make the decision to spend your money on an auxiliary speaker, see if your man at your local CB store will let you listen to one that has already been installed in someone's car. You may decide a speaker isn't necessary. On the other hand, if the requirements of your particular mobile installation have demanded that you mount the radio so its speaker is snuggled firmly against the car's dashboard or transmission tunnel, causing everything that comes out of it to be muffled, you may be a natural for one of these.

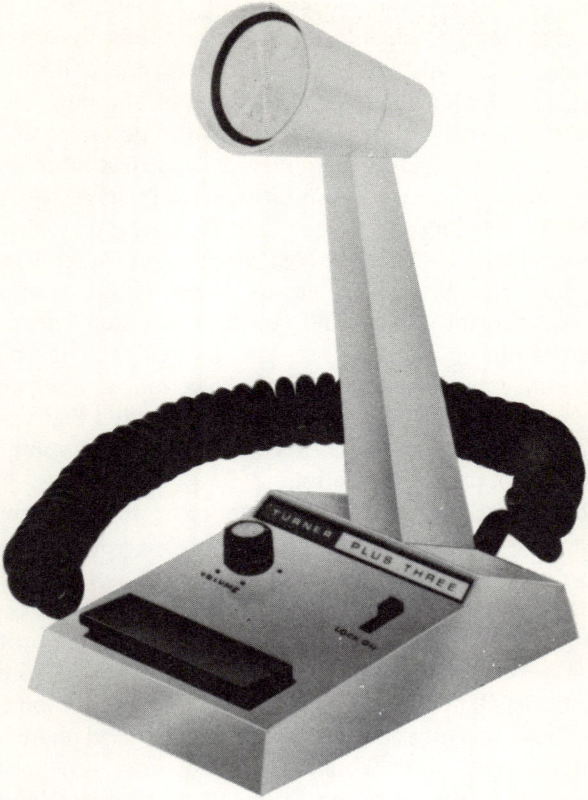

Figure 10-5. This microphone, used with a base station, incorporates a speech-compression circuit that is adjusted with the knob on the mike's base. Use of the circuit keeps the transmitter as close to 100 percent modulation at all times as possible. *Photo courtesy of Turner Division, Conrac Corporation.*

Another prime accessory that can go a long way toward improving the sound of your radio is an after-market microphone. These come in at least three different varieties—carbon, ceramic, and dynamic—because most radios come with a mike that uses one of these types of construction as standard equipment. When you buy an accessory mike you have to buy one that is of the same sort as your original mike. You also have to be careful to get a mike that is of the proper impedance, because this varies from radio to radio, too. Think for a minute about what a microphone does: It converts the air-pressure waves created by speech into electrical impulses that are sent to the transmitter and from there to the antenna and out over the airwaves. Obviously, it is a very important part of your system. An after-market microphone has the potential of making your voice tones sharper and clearer, and thus more readily understandable. In many instances you won't have too much choice about whether or not to use this accessory, because many radios come with their mikes wired as an integral part of the set; thus, changing mikes means a trip to the repair shop and the spending of money, and you probably don't need that. Some sets do have detachable mikes, though, and if yours is one of them, you're in luck. A large selection of after-market mikes is available, and it's possible to obtain units that will improve your set's performance, or control your set's volume, or maintain 100 percent modulation.

Another interesting device that can come in handy with a base station is a telephone patch, but before you get too involved with one of these the first thing to do is check with a representative of your telephone company to see what the company's attitude about phone patches is. A phone patch probably would be particularly useful if you became involved with one of the Citizens Band radio-service groups and were monitoring channel 9 and relaying emergency broadcasts. The phone company will be particularly concerned about how you hook the patch up, and may charge you for a telephone extension, or it may not be at all interested in the idea. If the phone company says no, just shrug your shoulders and forget the idea.

What a phone patch does is link Citizens Band radio communications to the telephone network. It allows you to link an incoming radio call to someone on the other end of your telephone line. They come in varying styles. Your basic telephone patch is just a little box with a knob, a switch, and some wires. Some of the wires you connect to a telephone terminal, and some you connect to the microphone and speaker terminals on your transceiver. When you get it all hooked up and throw the switch, the phone is connected to the radio, and vice versa. In most cases, when you use the patch you'll have to sit there and key the mike to make the transmitter work when the party on the phone is talking, and unkey it so that the party on the phone can hear the party coming in on your radio, but there is another sort of phone patch that relieves you of this. It is a cradle that you lay the telephone's handset into, and it's connected both to your home's telephone circuits and to your transceiver. When the party on the telephone speaks, a voice-actuated switch keys the transmitter; when the party on the telephone pauses, the station is automatically switched to receive, and whoever is on the far end of the radio waves can speak and be heard.

Suppose you live in an area with a lot of radio traffic, and you're interested in communicating only with a specific CB unit, one that may be part of your own little radio network. All the extraneous CB calls that constantly jam the radio airwaves can be a tremendous pain in the neck, so what you may need is a coded calling system. Most of these use a system called tone squelch, and with the system attached each radio that has a device hooked to it will emit a coded tone along with its modulated signal when the microphone is keyed. Some systems emit the tone just briefly when the mike is first keyed; others emit the tone for the duration of the transmission. A listening radio that is part of the system will be equipped with a decoder that turns on the radio's receiver section only when it hears the appropriate tone. It works well, and it keeps a good deal of the typical, idle CB conversation out of your ears, if that's what you want.

Then there is the speech compressor, some of which come already built into accessory microphones. It's designed to let your radio operate at as close to 100 percent modulation as possible. In most cases your modulation will hit 100 percent, its most effective level, only during certain voice peaks, at percussive noises made by your voice as you say words with hard letters in them like *P* and *K,* or when you speak particularly sharply to emphasize something. The rest of the time your modulation is likely to be a good deal lower than 100 percent. A speech compressor changes that, because it's designed to increase the strength of a soft input signal and decrease a strong one so that you have a fixed amount of input to your transmitter at all times. The overall effect is to increase the amount of audio another radio can hear you generating and make your communications more completely audible. Note that any device that boosts your set's modulation beyond 100 percent, even just for an instant, is patently illegal.

So now you've got it. If you've read this book all the way through, cover to cover, you have a fair understanding of how Citizens Band radio works and how you can get the best out of it. Reading this book has not made you an instant radio technician—that's not what it was designed for. But what you do know is how to buy the very best equipment, how to know just what you need to get started, how to install it all right the first time, and how to stay within the limits of the law. That being the case, if you've followed along carefully chances are your equipment will continue to offer top performance for a long time to come, your neighbors won't hate you for gumming up their television picture, and you'll be able to benefit from efficient radio communications. More than that you can't ask for.

APPENDIX A:
The 10-Code

Briefly put, the 10-code is verbal shorthand. Its purpose is to make radio communications easier and more understandable. Since most of the codes are briefer than the English words they replace, they take less time to utter. They enable you to spend less time on the air. Call it economy of speech.

Additionally, there are times when combinations of distance and outside interference can make it downright hard to hear what your contact is trying to say. In situations like this, using the code lets you concentrate on simple numbers instead of words, so instead of trying to make out a badly garbled "What's your location?" a clear, crisp "10-20?" gets the job done with a minimum of fuss and a fair chance of cutting through the muck.

There is one minor detail, though, that muddies up the water just a little: There's more than one 10-code. How come? That's a good question, but one for which nobody could really provide a satisfactory answer. And, unfortunately, the FCC has turned a blind eye to the problem, so no one code has its blessing.

Presented here is a traditional, long-form 10-code. There are some newer, shortened ones, and if you find one you like better, use it. Whichever code you decide you like, just remember that you must keep a copy of it in your logbook. A 10-code can save you considerable time and trouble if you learn it and use it properly.

10-1 Receiving poorly.
10-2 Receiving well.
10-3 Stop transmitting.
10-4 Affirmative, message received.
10-5 Relay message.
10-6 Busy, please stand by.
10-7 Out of service, going off the air.
10-8 In service, ready to receive calls.
10-9 Repeat the message.
10-10 Transmission completed, please stand by.
10-11 You're talking too rapidly.
10-12 I have visitors present.
10-13 Please advise about road and/or weather conditions.
10-16 Make a pick-up at ——.
10-17 Urgent business.
10-18 Do you have a message for me?
10-19 There is no message for you, return to base.
10-20 What's your location? (Or, My location is——.)
10-21 Call by telephone.
10-22 Report in person to ——.
10-23 Stand by.
10-24 I've completed my assignment.
10-25 Can you contact ——?
10-26 Disregard previous transmission.
10-27 I am moving to channel ——.
10-28 Please identify your station.
10-29 Your air-time limit is up.
10-30 Does not conform to FCC regulations.
10-32 I will give you a radio check.
10-33 This station is carrying emergency traffic.
10-34 I'm having trouble, I need help.
10-35 Confidential information.
10-36 Correct time. ("Can you give me a 10-36?")
10-37 Wrecker needed at ——.
10-38 Ambulance needed at ——.
10-39 Your message has been delivered.
10-41 Please turn to channel ——.
10-42 There is an automobile accident at ——.
10-43 There is traffic congestion at ——.
10-44 I have a message for you.
10-45 All units, please report to base.
10-50 Break channel ——.
10-60 The next message number is ——.

10-62 I'm unable to understand you, use a telephone.
10-63 Not directed to ——.
10-64 Not clear.
10-65 I am awaiting your next message/assignment.
10-67 All units comply.
10-70 There is a fire at ——.
10-71 Proceed with transmission in sequence.
10-73 There is a speed trap set up at ——.
10-75 You are causing interference.
10-77 Negative contact.
10-81 Reserve a hotel room for ——.
10-82 Reserve space for ——.
10-84 My telephone number is ——.
10-85 My address is ——.
10-89 A radio repairman is needed at ——.
10-90 I have TVI (TeleVision Interference).
10-91 Hold the microphone closer to your mouth.
10-92 Your transmitter is out of adjustment.
10-93 Check my frequency on this channel.
10-94 Please give me a long count.
10-95 Please transmit a dead carrier for five seconds.
10-99 Mission completed, all units secure.
10-200 The police are needed at ——.

Note: Most of the 10-code signals can be used either as questions or as statements of fact.

APPENDIX B:
A Dictionary of CB Radio Slang

Much of what you hear on CB radio these days is completely understandable, but there is still a considerable amount of language that is likely to be completely unfamiliar to you. For the most part it's truckers' lingo, a salty patois developed by over-the-road, long-haul truckers that is part slang, part code, and part the language of a growing subculture. It's interesting, colorful, and in constant use.

Here are a few of the most commonly heard terms and their meanings:

TERM	*COMMON USAGE*	*MEANING*
Back door	Will you watch my back door (keep an eye peeled for cops)?	Rear of a vehicle, last rig in a convoy.
Back her down	Back her down, Smokey's taking pictures.	Slow to the speed limit.
Barefoot	Are you running your radio barefoot?	A CB radio without a signal-boosting linear amplifier.
Break	Break, break for the Rubber Duck.	I wish to call on this channel.

TERM	*COMMON USAGE*	*MEANING*
Breaker	I'm reading you loud and clear, breaker.	Someone using the call channel.
By	We're by.	I'm standing by, listening, waiting.
Chicken coop	There's a chicken coop a mile west of town.	A weigh station for trucks.
Clear	We're clear.	We're off the air.
Come back	Break for the Rubber Duck, come back.	Acknowledge the call.
Come on	Break for the Rubber Duck, come on.	Acknowledge the call.
County Mounty	There's a County Mounty northbound on Fifth Street.	Sheriff's deputy.
Covered up	I can't hear you, you're covered up.	You're experiencing interference, your signal is not getting out.
Double nickels	Back her down to double nickels.	55 miles per hour, the national speed limit.
Ears	How 'bout it, Rubber Duck, you got your ears on?	A Citizens Band radio set.
Eighteen-wheeler	Any westbound eighteen-wheelers out there?	A truck-tractor-and-trailer combination.
Eights, or eighty-eights	Eights to you.	Good-bye, best of everything.
Feed the bears	I fed the bears on my last trip through here.	Pay a speeding fine.
Flip-flop	I'll give you a shout on the flip-flop.	The return journey of a two-way trip.
Foot-warmer	I'm running my rig with a foot-warmer.	A linear amplifier that boosts radio signal strength.
Four-wheeler	This is the blue four-wheeler at your back door.	A passenger car, van, or pickup.

TERM	*COMMON USAGE*	*MEANING*
Getting out	How am I getting out?	Being heard.
Gone	We gone.	End of transmission.
Green Stamps	Smokeys got some of my Green Stamps on my last trip.	Dollar bills.
Hammer	Put the hammer down.	Gas pedal.
Land line, or LL	Anybody got a land line?	Telephone.
Local yokel	Local yokel taking pictures ahead.	City policeman.
Negative contact	Negative contact on that Rubber Duck.	Unable to contact station being called.
Negatory	Negatory on that.	Slang word for negative.
Pull the plug	We gonna pull the plug.	Turn off the radio.
Plain wrapper	Smokey in a plain white wrapper ahead.	An unmarked patrol car.
Radio check	How about a radio check?	A check to make sure the radio is working properly.
Rig	What kind of rig you running?	Radio set.
Rocking chair	We're in the rocking chair.	In the middle of a convoy.
Shakey town	We're headed for Shakey town.	Los Angeles.
Skip	The skip is in.	Atmospheric condition that allows long-distance communication.
Smokey the Bear	Smokey the Bear is parked up ahead taking pictures.	State highway patrolman, used because some wear ranger hats.
Taking pictures	Smokey is up ahead taking pictures.	Using radar to catch speeders.

TERM	*COMMON USAGE*	*MEANING*
The big switch	We're gonna throw the big switch.	A radio's main power switch.
Thermos bottle	How about that west-bound thermos bottle?	A tank truck.
Walked on	Come back, we just got walked on.	Signal wiped out by a stronger signal.
We gone	10-4, we gone.	End of transmission.

APPENDIX C:
Mailing Addresses, Federal Communications Commission

As time goes by and you get more and more involved with Citizens Band radio, you may develop curiosities about how to get hold of a complete set of the FCC's CB rules and regulations (known as Volume VI, Part 95, which every CBer should have), about where to mail a license application, about how one moves from CB to amateur radio, or about areas of the FCC's activities not explored in this book.

Below you will find the addresses you need for mailing a license application, ordering the rules and regulations, and finding the addresses of your local common carrier field offices and field engineering offices.

If you're willing to wait for just a bit, a letter is probably the most effortless way of getting what you need, but if you're in a hurry and wish to visit or telephone your local FCC office, if there's one in your area, it will be listed under "United States Government" in your telephone directory.

The mailing addresses are as follows:

CB license applications
Federal Communications Commission
Gettysburg, Pa., 17325

Requests for CB rules and regulations
Superintendent of Documents
U.S. Government Printing Office
Washington, DC, 20402

(Don't forget to include your name and address and $5.35, the cost of the documents.)

APPENDIX D:
A Directory of Manufacturers

If you're wondering just how many manufacturers of Citizens Band radios and antennas there are, where they are, and who makes what—read on.

Radio manufacturers

Apelco
676 Island Pond Road
Manchester, NH 03103

Browning Laboratories, Inc.
1269 Union Ave.
Laconia, NH 03246

Courier
Fanon/Courier Corp.
990 S. Fair Oaks Ave.
Pasadena, CA 91005

Craig Corp.
921 W. Artesia Blvd.
Compton, CA 90220

Dynascan Corp.
1801 W. Belle Plaine Ave.
Chicago, IL 60613

E. F. Johnson Co.
Wauseca, MN 56093

Fieldmaster Radio Corp.
21212 Van Owen St.
Canoga Park, CA 91303

Handic of U.S.A.
14340 N.W. 60th St.
Miami Lakes, FL 33104

Hy-Gain Electronics Corp.
R.R. #3
Lincoln, NB 68505

J.I.L. Corp. of America
1000 E. Del Amo Blvd.
Carson, CA 90746

Kris, Inc.
Cedarburg, WI 53012

Lafayette Radio Electronics
Corp.
111 Jericho Turnpike
Syosset, NY, 11791

Midland International Corp.
P.O. Box 19032
Kansas City, MO 64141

Olson Electronics
260 S. Forge St.
Akron, OH 44327

Pace Communications
24049 S. Frampton Ave.
Harbor City, CA 90710

PAL Electronics
2962 W. Weldon
Phoenix, AZ 85017

Panasonic
One Panasonic Way
Secaucus, NJ 07094

Pearce-Simpson
P.O. Box 520800, Biscayne
Annex
Miami, FL 33152

Radio Shack
2617 W. 7th St.
Fort Worth, TX 76107

Ray Jefferson
Main and Cotton Streets
Philadelphia, PA 19127

Regency Electronics, Inc.
7707 Records St.
Indianapolis, IN 46226

Robyn International, Inc.
P.O. Box 478
Rockford, MI 49341

Royce Electronics Corp.
1142 Clay St.
North Kansas City, MO 64116

SBE Linear Systems, Inc.
220 Airport Blvd.
Watsonville, CA 95076

Sharp Electronics Corp.
10 Keystone Place
Paramus, NJ 07652

Sonar Radio Corp.
73 Wortman Ave.
Brooklyn, NY 11207

Sony Corp. of America
714 Fifth Ave.
New York, NY 10019

Standard Communications
108 W. Victoria St.
Carson, CA 90248

Teaberry Electronics Corp.
4655 Massachusetts Ave.
Indianapolis, IN 46218

Tram/Diamond Corp.
Lower Bay Road
Winnisquam, NH 03289

XTAL
9749 Shirley Ave.
Northridge, CA 91324

Antenna Manufacturers

Antenna, Inc.
23850 Commerce Park Road
Cleveland, OH 44122

Avanti Research & Development, Inc.
340 Stewart Ave.
Addison, IL 60101

Blue Streak
200 E. Wilshire Ave.
Santa Ana, CA 92705

Breaker Corp.
110 Great Southwest Parkway
Arlington, TX 76011

Cushcraft Corp.
621 Hayward St.
Manchester NH 03103

GC Electronics
400 S. Wyman St.
Rockford, IL 61101

Gotham
2051 N.W. Second Ave.
Miami, FL 33127

Hustler
New-Tronics Corp.
15800 Commerce Park Drive
Brook Park, OH 44142

Hy-Gain Electronics Corp.
R.R. #3
Lincoln, NB 68505

Lafayette Radio Electronics
Corp.
111 Jericho Turnpike
Syosset, NY 11791

Radio Shack
2617 W. 7th St.
Fort Worth, TX 76107

Robyn International, Inc.
P.O. Box 478
Rockford, MI 49341

Shakespeare Industrial
Fiberglass Div.
P.O. Drawer 246
Columbia, SC 29202

Turner Co., Inc.
909 17th St. N.E.
Cedar Rapids, IA 52402

INDEX

T

W